Vocal Projections

Vocal Projections

Voices in Documentary

Edited by

Annabelle Honess Roe and Maria Pramaggiore

BLOOMSBURY ACADEMIC
NEW YORK • LONDON • OXFORD • NEW DELHI • SYDNEY

BLOOMSBURY ACADEMIC
Bloomsbury Publishing Inc
1385 Broadway, New York, NY 10018, USA
50 Bedford Square, London, WC1B 3DP, UK

BLOOMSBURY, BLOOMSBURY ACADEMIC and the Diana logo are trademarks of Bloomsbury Publishing Plc

First published in the United States of America 2019

Cover design: Louise Dugdale
Cover image: Amy Winehouse, 2007 © Herve Allemand / ArenaPal

A catalog record for this book is available from the Library of Congress.

ISBN: HB: 978-1-5013-3125-1
ePDF: 978-1-5013-3127-5
eBook: 978-1-5013-3126-8

Typeset by Integra Software Services Pvt. Ltd.

Contents

List of Illustrations

Acknowledgements

This project was born out of a chance conversation in 2012 and the discovery of our unexpected mutual intellectual interest in the unacknowledged significance of the voice in documentary. This planted the idea for what would become *Vocal Projections: Voices in Documentary*, a project that has prevailed through many life, geographical and professional changes over the intervening years. As with many things, we think this book has benefitted from its slow gestation. It has allowed us to explore and play with ideas around the voice and documentary in various contexts, including a symposium we co-convened at the University of Surrey in 2014. The book has also developed in step with a growing interest in sound, and particularly the voice, in documentary studies. The publication of this book represents a rewarding fruition of our interest in this topic and, we hope, augurs well for future developments, as our work joins other recent publications in this area.

We are indebted to the contributors to this book, who were generous with their efforts and thoughts and very patient with our editing process. In addition, we have greatly enjoyed various intellectual exchanges about the voice in documentary with colleagues far and wide at venues such as the Society for Cinema and Media Studies and Visible Evidence conferences and, most recently, as part of the stimulating 'stream' on the documentary voice chaired by Pooja Rangan and Genevieve Yue at the 2017 American Comparative Literature Association conference in Utrecht.

Introduction: Documentary's Vocal Projections

Annabelle Honess Roe and Maria Pramaggiore

Vocal Projections: Voices in Documentary seeks to reconsider if not displace the primacy of the visual in documentary studies by focusing on the way documentaries engage with the sonic capacities and the semiotic implications of human and non-human voices. The vibrational, imaginative and political projections of voices, whose 'vague spatial anchoring' (Metz 1980: 29) enables them to be coupled with and decoupled from bodies and images, endow them with complex signifying possibilities. The key concerns of this book echo questions raised by Christian Metz in 'Aural Objects', where he contends that the Western understanding of sound as a secondary quality – 'a non-object' – leads to a 'tendency to neglect its own characteristics in favor of those of its corresponding "substance," which in this case is the visible object, which has emitted the sound' (1980: 30).

The scholarly disregard for the characteristics of sound in film and media studies that Metz identified in 1980 has been redressed in the ensuing decades by the work of Rick Altman, Claudia Gorbman, Anahid Kassabian, Michel Chion, Michael Bull and Les Back, Jean-Luc Nancy, Jay Beck and Tony Grajeda, Jonathan Sterne, Liz Greene and Danijela Kulezic-Wilson, among many others. Amid the 'sonic turn' that has informed film and media studies scholarship, however, only a few works devote their main attention to documentary sound, including Gunnar Iversen's and Jan Ketil Simonsen's *Beyond the Visual: Sound and Image in Ethnographic and Documentary Film* (2010) and Holly Rogers's *Music and Sound in Documentary Film* (2015), and only Pooja Rangan's and Genevieve Yue's recent special issue of *Discourse* (2017) specifically examines the voice.

Vocal Projections was prompted by what seemed to us to be this glaring and even inexplicable omission: although the concept of 'the voice' occupies a central position within documentary film and media studies, voices have never fully been given their due. Vocal metaphors dominate the field, having emerged from Bill Nichols's early discipline-defining essay, 'The Voice of Documentary' (1983), which frames the voice of documentary as a rhetorical sensibility that presents a film's 'social point of view' (1983: 18). In *Representing Reality* (1991) and *Speaking Truths with Film* (2016), Nichols continues to marshal metaphors that equate voice with argument and point of view. Nichols's most recent characterization of the voice of documentary is as a 'Hey You' that attempts to 'speak to us about the world we share' (2016: 72). However, the chapters in this book collectively argue that by examining the implications of documentary filmmakers and audiences treating the human voice as a 'substance' we can productively move beyond the conflation of voice and the sense-making imperatives of language that has dominated documentary studies. As such, each of the sixteen chapters in *Vocal Projections* considers the ways that documentaries attend to and exploit the attributes of voice by paying attention to the sonic principles, audio techniques and structural choices that filmmakers employ in their documentary treatment of recorded voices.

Nichols's conceptualization of the voice of documentary as a framework for delivering a social or political point of view comports with a long intellectual history beginning with Aristotle and continuing through Heidegger that understands the voice as an instrument of rational and intentional human communication. The notion that the voice offers access to a stable subjectivity has been questioned within post-structural and postmodern theory, notably by Roland Barthes, Jacques Derrida, Gilles Deleuze and Mladen Dolar. With the exception of Dolar, however, these theorists are more interested in challenging the coherence of the subject than in exploring the voice per se. Moving beyond ideas about individual identity, philosopher Adriana Cavarero and social theorist Nick Couldry highlight the voice as a site of political subjectivity, a means for asserting unique standpoints and establishing shared narratives that foster collective action.

Taking up Nichols's metaphorical conception of the voice, most studies of documentary have explored the voice in terms of its evidentiary, truth telling, rhetorical, persuasive and authenticating functions. The most significant challenge to Nichols's paradigm within documentary studies has emerged from postcolonial and feminist scholars, many of whom retain an understanding of the voice as a medium for political representation while mounting a critique of documentary's desire to speak for the underrepresented, the marginalized and the oppressed. Gayatri Spivak, Trinh T. Minh-ha, Fatimah Tobing Rony and Brenda Longfellow, among others, have indicted ethnographic documentary for reproducing colonial hierarchies, pointing to practices such as subtitling and dubbing that paternalistically objectify the people whose point of view documentary purports to be enabling. In particular, Trinh's influential concept of 'speaking nearby' has been taken up as an ethical intervention to reconfigure the power dynamics that inevitably result from the practice of speaking for others.

Vocal Projections does not seek to reject the fruitful insights that have emerged from the metaphorical conceptualization of the voice that treats it as a rhetorical and political instrument. Rather, we intend to augment this way of understanding the *voice of* documentary with an examination of *voices in* documentary. We believe that our shift in emphasis bears important implications. That voices are always specific rather than generic was implied by Asif Kapadia when asked why he chose not to use voice-over in *Amy* (2015). His response that 'I didn't want to use voiceover; I mean, who would I use?' (2016) suggests that this narrational technique is always informed by the acoustic effects and the social and political associations conveyed by a specific voice. Similarly, Stevan Riley's *Listen to Me Marlon* (2015), which draws from Marlon Brando's extensive private audio archive of interviews, home movies and informal verbal musings, foregrounds the complex relation between documentary voices and subjects. An exercise in volubility and immediacy, the film's intensely materialist engagement with Brando's voice reminds us of the limitations of ascribing a singular metaphorical understanding to documentary voice. Instead, as this book reflects, documentary's many voices are linked to both visible and implied bodies in a variety of ways and may serve to shore up or authenticate as well as to deconstruct identities.

More than simply acknowledging the fact that a range of diverse voices contributes to our experience of documentaries, we propose that engaging with vocal plurality – which encompasses not only the interactions among narrating, interviewed and overheard voices that are common to documentary but also the way that documentary subjects' 'own' voices are not always identical to themselves across contexts, times and places – provides opportunities for expanding our understanding of the formal organization and social functions of documentary. *Vocal Projections* attends to the way that documentaries mine voices for affective power, intersubjective potential and aesthetic appeal, often deploying them in ways that challenge their suitability as narrational mouthpieces that neutrally or transparently convey social perspectives.

If we dispense with the presumption that the voice is important primarily because it fulfills evidentiary and authenticating functions, we open up a range of possibilities for documentary. For example, we can consider how the use of recognizable voices to narrate commercial documentaries challenges the idea that the power of the 'Voice of God' comes from its unseen, unknown and disembodied status, as Annabelle Honess Roe argues in 'Playing God: Film Stars as Documentary Narrators', which opens Section One, 'Voices, Bodies and Power'. The three chapters in this section investigate the ways screen voices communicate social power through their connections to both visible and/or implied bodies. Pooja Rangan's 'Auditing the Call Centre Voice: Accented Speech and Listening in Sonali Gulati's *Nalini by Day, Nancy by Night* (2005)' considers the complex dynamic of giving image to the voices of Indian call centre agents, who are rendered faceless and accentless by design. In 'The Fundamental Lie: Lip Sync, Dubbing, Ventriloquism and the Othering of Voice in Documentary Media', Patrik Sjöberg examines the uncanny mismatch of voice and body in experimental videos by Adie Russell and Gillian Wearing, arguing that their work elicits new ways to negotiate a matrix of visible and implied bodies in non-fiction media.

The four chapters in Section Two, 'Voices beyond Language', consider documentary voices that exceed or refuse semantic logic and, as a result, dissolve the boundaries of subjectivity. Maria Pramaggiore's 'Hearing Voices in Singer-Songwriter Music Documentaries' explores archivally rich singer-songwriter films that promise enhanced, immersive access to the multiple voices of musicians, yet paradoxically resist the viewer's ability to discern the 'truth' behind the voices. In '"The Voice" and Sound in Indian Documentary Film: Listening to Surabhi Sharma's *Bidesia in Bambai* (2013)', Deborah Matzner argues that song invokes several modes of engagement by mediating affect within the film while also celebrating the tenacity of migrant communities whose members identify with the experience of dislocation through non-linguistic vocal expression. Shilyh Warren's 'Hum, Buzz, Gurgle: Ecological Soundscapes in Poetic Ecodocs' examines ecological documentaries that employ vocal alternatives to the human voice – including those that emanate from animals, water and machines – to produce a sense of wonder and enchantment, rather than fear and pessimism, in the face of challenges wrought by the Anthopocene. In 'Literally Documenting Her Voice: Valie Export's Vocal-Visual Experiments with Screen Media', Helen Hughes concludes this section with a study of a feminist film, video and installation artist whose interests lie in deconstructing 'rebellious' voices.

Export's vocal dissonances dissolve language and gender difference as they radically reject the idea that any voice reflects a fixed identity.

Section Three, 'Gender, Sexuality and Species', draws our attention to documentary voices that are linked to alternative identities and reveals the challenges they can pose for film form, genre convention and spectatorship. Sarah Kessler's 'The Voice of Mockumentary' examines ventriloquism in the mockumentary, analysing the way the 'gay voice' in *Waiting for Guffman* (Guest 1997) and *Modern Family* (ABC 2009–ongoing) exposes documentary's contradictory authentication practices for alternative social subjects. In 'Fabulous Thinking, Chimeric Voice: Kathy High's *Lily Does Derrida: A Dog's Video* (2010)', Brenda Hollweg probes the manner in which vocal fabrication enables Kathy High to employ the cine-essay form to develop democratic notions of trans-species communication. Jean Walton's 'Animating Voices, Onscreen and Off, in Kathleen Shannon's *Working Mothers* (1974–75)' details the way Kathleen Shannon and Studio D, the Canadian studio devoted to making films by and about women, focused on women's voices on screen and off as a means of integrating feminist film practice, engaged spectatorship and consciousness raising.

The three chapters in Section Four, 'Documentary Modes and Techniques', address the distinctive vocal aesthetics of recent documentaries within industry and cultural contexts. Chris Cagle's 'The Voice in Documentary Sound Design: A Digital Revolution' identifies the ways digital sound design enhances the expressive possibilities of environmental sound in contemporary documentary. In the process, Cagle contends, they renegotiate the relationship between voice and environment, producing three recognizable forms of vocality: the spatialized voice, the spectral voice and the uncanny voice. In 'Kazuhiro Soda's Migrant Voices', Luke Robinson argues that Japanese director Soda's *Oyster Factory* (2015) offers viewers a way of understanding the experiences of Chinese migrants in Japan without trying to 'speak for' those subjects. Soda combines strategic subtitling with standard observational techniques to imply that 'giving voice' itself is a socially constructed process. Informed by his work as a documentary practitioner, James Nicholson's chapter, 'Vocal Hierarchy in Documentary', concludes this section. Nicholson asserts that that vocal power is expressed relationally and he incorporates several contemporary documentaries into his analysis.

The three chapters in the concluding section, 'Time, Testimony and History', assess the confessional, testimonial and memorial functions of voices. The competing claims which cast the voice as both the source of irrefutable evidence and an unreliable medium were driven home in Andrew Jarecki's *The Jinx* (2015) and Laura Ricciardi's and Moira Demos's series *Making a Murderer* (2015–16), two texts explored by Kristen Fuhs in 'Crime Documentary's Confessing Voice'. Fuhs argues that the aural qualities of and context within which the confessing voice is heard are just as significant for interpretation as the words that have been spoken. Ohad Landesman's 'Sounds of Disillusionment and Discord: When Pasolini and Sontag Visited Israel', probes the way these two filmmakers use their own voices to establish their outsider status and reveal their disillusionment at the discrepancies between an imagined and a real Israel. Rhiannon Harries's 'Beyond Words: Speech, Ethical Time and the Politics of the Local in Nicolas Philibert's *Retour en Normandie* (2007)' concludes this final section with the contention that Philibert's employment and elision of voice, speech and language challenges traditional notions of temporality in historical documentary.

Vocal Projections: Voices in Documentary explores the way that voices may construct but also challenge and defy subjective and narrative coherence, audiovisual form and documentary conventions. This collection thus takes up the diverse uses and meanings of the voice not merely as a platform for the dissemination of language, rhetoric or story, but also as extra-linguistic expression that is just as likely to undermine truth claims as to uphold them. By going beyond the conception of 'the voice' of documentary as reflecting a social stance or point of view, this book suggests that the attributes and affordances of the human voice are just as important as, if not more important than, the rational discourse for which the voice can also serve as a vehicle.

References

Cavarero, A. (2005), *For More than One Voice: Toward a Philosophy of Vocal Expression*, Palo Alto, CA: Stanford University Press.

Chion, M. (1994), *Audio-Vision: Sound on Screen*, trans. C. Gorbman, New York: Columbia University Press.

Chion, M. (1999), *The Voice in Cinema*, trans. C. Gorbman, New York: Columbia University Press.

Chion, M. (2016), *Words on Screen*, trans. C. Gorbman, New York: Columbia University Press.

Couldry, N. (2010), *Why Voice Matters: Culture and Politics after Neoliberalism*, London: Sage Publications.

Dolar, M. (2006), *A Voice and Nothing More*, Cambridge, MA: MIT Press.

Iverson, G. and J. K. Simonsen, eds (2010), *Beyond the Visual: Sound and Image in Ethnographic and Documentary Film*, Højbjerg: Intervention Press.

Kapadia, A. (2016), 'Talking Documentary', Galway Film Centre workshop, Harbour Hotel, 18 January 2016.

Kramer, L. (2014), 'On Voice: An Introduction', in W. Bernhart and L. Kramer (eds), *On Voice*, vii–xv, Boston, MA: Rodopi/Brill.

Longfelllow, B. (2004), 'The Great Dance – Translating the Foreign in Ethnographic Film', in A. Egoyan and I. Balfour (eds), *Subtitles: On the Foreignness of Film*, 337–353, Cambridge, MA: MIT Press.

Metz, C. (1980), 'Aural Objects', trans. G. Gurriero, *Yale French Studies* 60: 24–32.

Nichols, B. (1983), 'The Voice of Documentary', *Film Quarterly* 36 (6): 17–30.

Nichols, B. (1991), *Representing Reality*, Bloomington, IN: Indiana University Press.

Nichols, B. (2016), *Speaking Truths with Film: Evidence, Ethics, Politics in Documentary*, Oakland, CA: University of California Press.

Rangan, P. (2017), 'Audibilities: Voice and Listening in the Penumbra of Documentary: An Introduction', *Discourse* 39 (3): 279–291.

Rogers, H., ed. (2015), *Music and Sound in Documentary Film*, London: Routledge.

Rony, F. T. (1996), *The Third Eye: Race, Cinema and Ethnographic Spectacle*, Durham, NC: Duke University Press.

Spivak, G. C. (1988), *Can the Subaltern Speak?*, Basingstoke: Macmillan.

Trinh, T. M. (1991), *When the Moon Waxes Red*, New York: Routledge.

Section One

Voices, Bodies and Power

1

Playing God: Film Stars as Documentary Narrators

Annabelle Honess Roe

Over the past few decades the use of well-known film actors to narrate mainstream, commercial documentary films and television series has become increasingly common. We can think of Tom Cruise's enthusiastic and awe-inspired narration of the 2002 IMAX documentary *Space Station*, John Hurt's mellifluous intonation accompanying the BBC's 2011 natural history series, *Planet Dinosaur* and, perhaps most famously, Morgan Freeman's anthropomorphizing voice-over for 2005's *March of the Penguins*. The use of successful thespians to narrate documentaries is not unprecedented; Orson Welles's voice-over for the 1958 ethnographic travelogue *Masters of the Congo Jungle* and Laurence Olivier's role as the narrator of the 1973 British twenty-six-part TV series *World at War* are two early examples of this strategy.

However, since around 2000, documentary has witnessed the growth of the star or celebrity narrator,[1] concomitant with the rise of documentary's commercial viability in the theatrical marketplace and its status as 'prestige' fare on television. This development, as this chapter will argue, challenges the previously accepted idea that the power of the documentary 'voice-of-God' narration rests with its status as a disembodied, anonymous voice. Instead, the power of the new documentary narrator comes as much from the specific body to which the voice belongs and includes the intertextual baggage that voice carries with it from other roles and off-screen activities. In addition, the rise of the star documentary narrator is indicative of the shift in cultural power towards the individual, and the celebrity in particular, and away from the general governing institution of the state represented by the historical voice-of-God.

The voice-of-God in documentary

First used extensively in the Griersonian documentaries made in Britain from the 1930s onwards, and a key feature of what Bill Nichols has termed the expository mode of documentary, the documentary voice-over in the form of voice-of-God narration is most commonly understood as a device of instruction; the unseen, but all-seeing, off-screen narrator guides the audience as to the meaning of the images it accompanies and how they should be interpreted. For example, in 1935's *Coal Face* (Alberto Cavalcanti), which was produced by the Grierson-headed GPO Film Unit, the voice-over acts as reinforcement and clarification of the images. The viewing guidance remains straightforward and factual, such as when information about regional mining productivity is heard over images of an illustrated map of the UK ('Durham and Northumberland: 146,000 men; 46 million tonnes').

Other films from this period, however, use the voice-over to more explicitly guide the viewer to see the images from a particular point of view. In *To-Day We Live* (Ruby Grierson and Ralph Bond 1937) the voice-over establishes the historic decimation of rural communities through a chain of causal events – the Industrial Revolution and urban growth, the First World War, the Great Depression – that is not apparent in the images themselves. Instead, it requires the narrator to interpret for us a shot of a row of terraced houses in an unnamed city by stating that the urban communities that grew up around new areas of industry had 'nothing in common except the sameness of landscape and an urge to work'. This urban alienation is contrasted with the 'intimacy of these cottages' seen in the film's opening scenes of rural village life, where 'villagers dwelt, quietly pursuing their labour on the surrounding land'. It is in the voice-over in *To-Day We Live*, far more than the images, that the film makes the argument that the loss of rural communities is something to be mourned. This then gives resonance to the film's subsequent illustration of the positive contribution being made to the rebuilding of such communities through activities supported by the National Council of Social Service, which sponsored the film.[2] In both these examples of 1930s British documentaries, sound, and specifically the spoken voice, dominates and the expository voice-of-God speaks for the images.

Similarly, in the United States, the narration heard in the *March of Time* newsreels screened in movie theatres in the 1930s and 1940s 'often serves as the prototype [for the voice-of-God]: stentorian, aggressive, assuming a power to speak the truth of the filmic text, to hold captive through verbal caption what the spectator sees' (Wolfe 1997: 149). Here Wolfe points to why, generally speaking, the voice-of-God is a maligned feature of documentary film. Emile de Antonio went so far as to describe it as 'fascist' because, as Keith Beattie (2008: 13) puts it, this voice 'limits and controls the possibilities for spectatorial interpretation'. This distrust of the voice-of-God extends to documentary scholars who, as Stella Bruzzi (2006: 57) points out, demonize the voice-over because it 'is assumed to be undemocratic and inherently distortive. There is, therefore, the suspicion that a voice-over has the capacity to violate the "truth" revealed in the image'. For example, the damning description of urban life in *To-Day We Live*, which equates images of life in an industrial city with social disenfranchisement, denies potential readings of any positive aspects of living in such environments.[3]

The acknowledgement of the limitations of this type of authoritative voice-over meant that it fell increasingly out of favour with documentary makers after the Second World War. The subsequent elevation of the image over pre-scripted words to convey a documentary's point of view, a move championed by Direct Cinema, meant that didactic narration was relegated to the staid and unadventurous realms of television current affairs. This was particularly the case in the United States, where the voice-of-God became an 'ethical guarantor' in the narration of Edward R. Murrow and his contemporaries in 1960s Network television documentaries (Renov 2004: 19). The voice-over was then 'rediscovered', according to Jeffrey Ruoff (1992: 222), in the late 1980s by documentary makers such as Michael Moore and Jill Godmilow, who used 'personal, ironic and interpretive commentary to counterpoint the synchronous images and sounds'. These filmmaker voice-overs can be thought of as a type of hybrid between the subjective, reflective narration heard in personal diary films, such as the work of Jonas Mekas in the 1970s, and the conventional voice-of-God. While Michael Moore's narration, for example, is personal, it is also strongly guiding us on how to interpret the images in his films. Documentary scholarship has recognized this historical diversification of documentary narration beyond the voice-of-God model established in the 1930s (see, for example, Bruzzi 2006). The rise of celebrity narrator, however, points to a resurgence of the voice-of-God in documentary, albeit a different God to that of the classical era of the 1930s and 1940s.

The power of the disembodied voice

The voice-of-God narrates from 'a position of absolute mastery and knowledge outside the spatial and temporal boundaries of the social world the film depicts' (Wolfe 1997: 149). This mastery over the images, the documentary subjects, the narrative and the viewer has been understood as coming from the anonymity.and omniscience of the narrator, characteristics that are, as Keith Beattie (2008: 12) points out, 'derived from, and seated in, the disembodied voice'. This echoes Mary Ann Doane (1980: 42), who suggests that the documentary voice-over becomes what she calls a 'guarantee of knowledge [...] precisely because the voice is not localizable, because it cannot be yoked to a body'. Thus, in examples such as *Coal Face* and *To-Day We Live* the narrators are situated in some other place beyond and above the image, their bodies remaining unseen – elsewhere and yet at the same time everywhere.

Here we might consider whether a formulation that hangs heavily over many discussions of off-screen voice is useful in helping to understand the power of the disembodied documentary voice-over. That disembodied, non-localizable voices that exist outside the current space of a film's diegesis are powerful is an idea central to Michel Chion's concept of the *acousmatic* voice – a voice that comes from 'an immaterial and non-localizable body' (Chion 1999: 24). The *acousmêtre* is 'a special kind of being' that is created when we hear a voice 'that has not yet been visualized, that is, when we cannot yet connect it to a face' (Chion 1999: 21). On first consideration, the *acousmêtre* would seem to have much in common with the documentary voice-of-God, particularly so when Chion (1999: 24) describes the *acousmêtre* as having a fourfold power: 'the ability to be everywhere, to know all, and to have complete power. In other words: ubiquity, panopticism, omniscience and omnipotence'. In his most detailed discussion of the *acousmêtre*, in *The Voice in Cinema*, Chion remains somewhat vague, however, as to whether voice-over narration is *acousmatic*. Initially, he seems to include this type of voice in his discussion when he delineates three types of *acousmêtre*: the complete *acousmêtre* ('one who is not-yet-seen, but who remains liable to appear in the visual field at any moment'); the already visualized *acousmêtre* (who 'you continue to hear after she leaves the visual field'); and 'the commentator-acousmêtre, he who never shows himself but has no personal stake in the image' (Chion 1999: 21). Chion's 'commentator-acousmêtre' seems to fit the description of the voice-over narrator in documentary film – a speaker who is never seen and does not insert himself or herself into the image. However, shortly after this tripartite definition, Chion seems to contradict himself by saying that the *acousmêtre* does not 'merely speak as an observer (as commentary)' (23) and that 'the acousmêtre, as we have noted, cannot occupy the removed position of the commentator, the voice of the magic lantern show' (24). In a subsequent book, Chion is more definitive, defining a voice as *acousmatic* if 'the bearer of this voice is theoretically capable of appearing onscreen at any moment' and clarifying that 'the acousmetre is distinct from the voice-over that is clearly external to the image' (Chion 2009: 466). It would seem then that the voice-of-God is not an *acousemêtre*.

In fact, we could go further and argue that it would be impossible for a documentary voice-of-God to be an *acousmêtre* because the *acousmêtre*'s power is vested in its relationship to the fictional space of the film. Indeed, Chion's discussion of the power of the off-screen voice in cinema is limited to narrative, fiction film and he makes no direct mention of non-fiction cinema or documentary. For Chion, the *acousmêtre* is powerful because it is the off-screen voice of a character whose body has previously appeared (or might yet do so). Like the Wizard in *The Wizard of Oz* or Mother in *Psycho*, such a voice *is* yoked to a body: one that is currently unseen but threatens to invade the image, and therefore the narrative, at any time. In films such as *The Wizard of Oz* and *Psycho* the *acousmêtre* retains its power only for as long as it is unseen; as soon as the voice is yoked to a body and located within the frame, it is divested of its powers – the power of the all-seeing, unseen voice (Chion 1999: 28–29).

Even though the documentary voice-of-God cannot be defined as an *acousmatic* voice, can Chion's formulation of the *acousmêtre* still help us understand the power of this type of documentary narration? Chion's suggestion that the *acousmêtre* ceases to be powerful once it is embodied seems to reinforce how documentary scholars have thought of the voice-of-God – a voice that is powerful precisely because it is disembodied, anonymous and outside of the space of the image. However, I will argue that while star documentary narrators remain disembodied for the duration of the films they narrate, they are far from unknown. Unlike the divestiture of power that takes place when the *acousmêtre* is embodied, 'dooming [it] to the fate of ordinary mortals' (Chion 1999: 29), the association of star documentary narrators with a specific body is far from disempowering. In fact, I suggest, the extra- and intertextual identities of celebrity narrators only enhances their omnipotence. Far from mere mortals, these are the voices of those who are already identified as occupying a rarefied status in comparison to the rest of us.

The quality of the voice

In addition to the narrator's localization in off-screen space, there is also something specific about the quality of the documentary voice-of-God that lends it authority. That is, *how* the words of the voice-over are spoken is as significant as *what* those words are. Chion has observed, talking more generally about the voice, the tendency of listeners to 'from the speech act [...] usually retain only the signification it bears, forgetting the medium of the voice itself' (Chion 1999: 1). This means that the expressive and communicative power of the voice itself gets forgotten because we tend to conflate it with speech. Mladen Dolar stresses the distinction between voice and speech when he says of the voice:

> It is *what does not contribute to making sense* ... It is there, in the very act of saying, but it eludes any pinning down, to the point where we could maintain that it is the non-linguistic, the extralinguistic element which enables speech phenomena, but cannot itself be discerned by linguistics. (Dolar 2006: 15, emphasis in original)

Chion and Dolar are here drawing attention to the fact that voice is something that goes beyond the language we hear it speak; the voice gains supplemental meaning through tone, delivery and timbre. This is a significant observation in terms of understanding how certain types of voices have become associated with the voice-of-God. It is no accident, therefore, that conventionally, and also problematically, the authoritative documentary voice-of-God is typified by a well-spoken and relatively deep male voice.

This 'norm' for documentary voice-over was established, in the UK, in the Griersonian documentaries, where voice-of-God narrators spoke in Received Pronunciation (RP), the plumby British accent favoured by the BBC from the 1920s onwards and henceforth accepted as the appropriate way to deliver recorded speech (British Library ND). The supposedly accentless vocal delivery of RP that betrayed nothing of its speaker's geographical origin, but much about their class and educational background, can be used to describe Olivier's speaking voice in *World at War*, while Welles's accent in *Masters of the Congo Jungle* is exemplary of its US equivalent: mid-Atlantic English.[4] The distinction between RP and more everyday vernacular is emphasized in films such as *Housing Problems* (Arthur Elton and Edgar Antsy 1935), produced by the GPO Film Unit and sponsored by the British Commercial Gas Association. This film includes other voices in addition to the unseen (and unnamed and uncredited) narrator. Here the clipped tones of the voice-of-God contrast with the local accents of both the slum-dwelling subjects of the documentary that feature in on-screen interviews and also the town councillor who is heard (but not seen) talking on the soundtrack. Unlike the film's narrator, however, Councillor Lawder is introduced by name ('first, Councillor Lawder, Chairman of the Stepney Housing Association will tell you something of the problem of slum clearance').

Although *Housing Problems* was groundbreaking for its inclusion of the voices of its working-class documentary subjects, it still maintains a clear power differential between the voice-of-God, Lawder and the slum tenants. This is achieved in part through the sound editing: the voice-of-God presides and introduces Lawder and the interviewees. But also, while Lawder's vocal intonation is clear and measured, it betrays traces of regional identity not heard in the film narrator's voice. Further distinction comes in the voices of the slum tenants, whose accents more clearly indicate class, educational background and geographic origin. Thus, it is from the particular sound and vocal qualities of the different voices, as much as through the editing, that *Housing Problems* establishes the relative power of the different speakers.

The interplay between the quality of the human voice and the language we hear it speak is what Roland Barthes termed the 'grain' of the voice (Barthes 1977). Although Barthes was talking about the singing voice in his essay on this topic, his ideas translate readily to a consideration of why certain actors narrate certain documentaries. We might think, for example, of Kenneth Branagh's narration for the 1999 BBC natural history documentary series *Walking with Dinosaurs*, which reconstructed prehistoric life using computer animation and animatronics. The series was at the vanguard of using computer animation in natural history and science programming to create photorealistic imagery and it gained much praise, at the time, for how believable the dinosaurs looked. This photorealism is enhanced by the way the series copies the camerawork and mise en scène of nature documentaries – similar camera angles, pacing and so on. Yet, of course, the images in the series are not 'real' – they were created digitally rather than captured on a camera.[5] In addition, a considerable amount of the content of the series was, in fact, hypothetical – especially the claims regarding dinosaur behaviour and habits. This is where Branagh's voice-over becomes particularly important. The script that Branagh reads makes no acknowledgement of this hypothetical status of many of its assertions and, in addition, speaks as if the images are in fact filmed images of actual dinosaurs. He always talks in the present tense and colludes in the ontological obfuscation of the image's origin. For example, in episode 5 ('Spirit of the Ice Forest') Branagh says 'under the sleeping trees it is almost pitch black, but with image-enhancement it is possible to get a laellynasauras's eye-view of the clan' as the image changes from a black screen, on which we can see nothing, to a green-tinged, grainy monochrome depiction of a group of diminutive dinosaurs foraging under arctic forest ferns. Of course, image enhancement was used to create this shot, but not in the way that Branagh's voice-over implies, that is, the use of 'night vision'-style filming technology that would enable a natural history film crew to shoot under the cover of darkness.

The ontological and factual elisions in Branagh's script are reinforced by the way he speaks the words of the script. His vocal delivery has much in common with RP and effectively establishes his status as an authoritative and omniscient voice-of-God. In addition, the quality and tone of his voice – a deep, warm and imposing baritone – help to dramatize the narrative of each of the six episodes of the series. Branagh emphasizes key points in the story, his resonant voice rising and falling in step with the events on screen – deepening at moments of impending doom and culminating in triumphant exclamations that correspond to climaxes in the prehistoric narrative. At the beginning of the first episode Branagh narrates a scene in which a coelophysis drinks from a river. Having established the dinosaur as 'light-boned, fast ... ' Branagh's voice then changes tone as he says 'she is also beautifully adapted ... for killing'. The dramatic pause before 'for killing' combined with Branagh's vocal shift from a more neutral, informative tone of voice that is evenly paced to one that is deeper and more emphatic is enhanced by a background score of glissando, crescendoing strings that reach their peak as the dinosaur snaps a fish out of the river with lightning speed. The palaeontological record probably does provide robust factual evidence for the claims made by Branagh in this example. Yet, other claims that are more tenuous are delivered in an identical fashion. Just a few seconds after the coelophysis displays her hunting skills we are told that she is frightened away by the arrival of a herd of placerias, something that is signalled by 'a growing chorus of calls'. However, what these bull-like dinosaurs sounded like, and indeed whether they could 'chorus' is conjecture and the lack of geological evidence of what dinosaurs sounded like is glossed over by Branagh's narration.[6]

The confidence of Branagh's vocal delivery in *Walking with Dinosaurs* received some criticism, with Nancy Banks-Smith (1999) in the *Guardian* saying it was 'surer of itself than the facts warranted'. In a different context,

Leslie Chow (2006: 35) has criticized Branagh as someone who 'overtastes his lines – he uses so much nuance that every syllable is over-stuffed and there is perhaps too much emphasis on resonance at the expense of meaning'. However, it is precisely Branagh's confident and dramatic delivery, paired with the script's brazen disregard for the contingent nature of the moving images it narrates, that makes it effective. Here, *Walking with Dinosaurs* evidences what Bruzzi has ascribed as a key feature of voice-of-God narration – a 'constant perspective [...] to which all elements within the film conform' (2006: 49). As Bruzzi goes on to point out, this is an important distinguishing feature between a voice-of-God narrator and the 'inner monologue' style of voice-over narration that is heard in poetic and personal documentary filmmaking. The voice-of-God narration cannot be heard to 'falter' because this would 'acknowledge[s] its inadequacy' in a way that would allow the 'personal, subjective potential' of the voice to surface and disrupt the veneer of omniscience and omnipotence (2006: 63).

Similarly, in the 2005 documentary *March of the Penguins* the quality of Morgan Freeman's voice works to underscore the film's themes. Here, Freeman's voice, which manages to be warm and bright as well as powerful and soft, helps evoke empathy with the documentary's subjects – emperor penguins battling against the harsh environment of the South Pole to procreate. His voice underscores the script's anthropomorphism and its narrativization of the penguin's mating cycle and Freeman sets the tenor of the film during the opening credits when he says 'the average temperature here at the bottom of the earth is a balmy 58-degrees below ... that's when the sun is out'. Such language is typical of the script, which is colloquial and empathetic. It also places much emphasis on the penguins' supposed emotional response to their situation. At one point, for example, we are told 'the loss is unbearable' as we see an adult penguin nudging at the frozen body of its young chick that perished in a bitter snowstorm. The script is also shaped around a narrative trajectory that positions the 'love story' to which the film bears witness as part of a longer history of this inhospitable region in which the penguins are 'one tribe' that 'legend has it, stayed behind' after the geological changes that turned the South Pole from 'densely forested' tropics to ice and saw other inhabitants depart 'long ago'.

Even more effective, or indeed affective, in creating the uplifting tone of the film is Freeman's folksy delivery of the words. 'Temperature' is 'temp-er-rat-CHEWER' and his southern accent and gentle enunciation of words such as 'balmy' ('baa-hlmy') are a sign that this is not going to be a hard-nosed or scientifically robust exploration of the hardships of a life lived in a sub-zero climate. Instead, Freeman's lilting tone indicates the heart-warming and at times amusing approach the film takes to documenting the natural world. Unlike Branagh, Freeman's voice is perhaps atypical of the traditional voice-of-God. His accent, in betraying his origins in the Southern United States,[7] is far from mid-Atlantic English and it lacks the more formal, 'thespian' vocal delivery of an actor such as Branagh. Importantly for this film, however, Freeman 'provides warmth with his typically comforting voice' (Lemire 2005) – a voice that has been 'universally recognised as one of the greatest voices of all time'[8] (Blair 2016). In *March of the Penguins*, the effectiveness of Freeman's voice-over does not derive from the authority associated with accents such as RP or mid-Atlantic English, but more from the particular qualities of his voice that have led it to be publicly deemed as 'beautiful' (Blair 2016) and 'avuncular' ('DS' 2005), and thus a suitable match for the metaphorical tone of voice of this documentary. Importantly, however, despite the personable qualities of Freeman's voice the voice-over is not subjective, tells us nothing of Freeman himself and still offers the 'constant perspective' typical of the omniscient voice-of-God (Bruzzi 2006: 63).

Equally important, Freeman's voice is 'that of a dominant, strong male figure' due to its depth and resonance (Oakland 2016), much like Branagh's. The sonorous quality of Branagh's and Freeman's 'robust' voices, as David Sonnenschein (2001:145) would describe them, is something Bruzzi (2006: 64) has observed as a 'crucial component' of the traditional documentary voice-over with its 'tones of authority and universality'. As Mary Ann Doane (1980: 42) points out, the possession of knowledge and the privileged position of interpretation are the preserve of the male speaker. While the accents of the voices heard narrating documentary film have evolved beyond the RP and mid-Atlantic English that dominated prior to the Second World War, it is still relatively unusual to hear a female voice-of-God. The exclusion of women from the role of documentary narrator is rooted in a much longer history of recorded media and beyond, such that 'the invention of the megaphone, loudspeaker, and microphone did nothing to change the common belief that women made poor orators because their voices weren't powerful enough' (Karpf 2006: n.p). Such historic prejudice against women narrators has led to the female voice being associated with specificity, according to Stella Bruzzi (2006: 59), in contrast to the universality of the male voice that affords it the status of the voice-of-God.

There are some notable exceptions. For example, Tilda Swinton has voiced TV and feature documentaries, including *Climate of Change* (Brian Hill 2010) and

Galápagos (BBC and National Geographic 2006) and Sigourney Weaver lent her voice to the US broadcast of the BBC natural history series *Plant Earth* (2007) replacing the voice of David Attenborough for an audience less familiar with his iconic vocal association with the genre in the UK.[9] However, both Swinton and Weaver are in possession of relatively deep voices without the 'sultry' tones associated with the equally low vocal registers of an actress such as Scarlett Johansson (McCluskey 2015). As such, Weaver and Swinton avoid the 'Catch 22' that 'high voice in women was associated with demureness, and low voice with sexuality, so that [...] the voice that escaped accusations of promiscuity wasn't considered authoritative enough for serious broadcasting' (Karpf 2006: n.p.). Weaver and Swinton, and more recently Daisy Ridley (better known for playing Rey in the 2015 *Star Wars* film *The Force Awakens*) in the 2016 documentary *The Eagle Huntress* (Otto Bell), are unusual in that their casting as the voice-of-God suggests their voices are deep enough to be authoritative without being deemed 'too sexy'. As such, even though these female narrators may seem to challenge the equation of the male voice with the voice-of-God, their vocal androgyny suggests otherwise and as Bruzzi (2006: 66) notes 'women narrators in mainstream film and television conform more readily to the masculine voice-over model'. In the case of Swinton, Weaver and Ridley, this vocal androgyny reflects their on-screen identities. Swinton and Weaver in particular have made careers playing gender-bending roles: Weaver is most famous for her role as Ripley in the *Alien* franchise, a role written for a male actor, and Swinton first came to wider public attention in Sally Potter's *Orlando* (1992) in which she played the gender-fluid titular character. Their subsequent casting as documentary narrators implies that women who can adopt the physical and vocal traits of masculinity are more likely to be permitted to assume the mantle of the voice-of-God.

Vocal identity

So far, the discussion of the voices of star voice-of-God narrators has illustrated the extent to which they conform to the dominant understanding of how this type of documentary narration establishes its power over the images and over the viewer. However, in one important regard they are different: they are not anonymous voices that cannot be 'yoked' to a body, as Mary Ann Doane put it. Nor are these voices 'absent personality', as Bruzzi (2006: 63) has described the voice-of-God. Instead, the unique qualities of their individual voices, something that is a 'distinctive and defining feature' of their 'star persona' (Shingler 2012: 80), very much conjure up the specific body to which it belongs. This is because, Anne Karpf (2006: n.p) notes, 'timbre is idiosyncratically, biologically controlled' and is why 'actors almost always remain recognisable: whatever role they assume'. This coupled with human beings' advanced capabilities in voice recognition (see Kreiman & Sidtis 2011: 176) means that in the case of *March of the Penguins* and *Walking with Dinosaurs*, and many more documentaries besides, the unseen, off-screen voice *can* be yoked to a body because we recognize the voice of the narrator. And the shadow of those bodies hang intentionally and significantly over the documentaries and reinforce further the power and authority of the unseen (but in these cases not unknown) voice-of-God.

Just as Swinton and Weaver's androgyny has become part of star personas that permit them to occupy the position of the typically male voice-of-God, Branagh and Freeman were cast in their narrator roles not only, one can presume, for the quality of their voices and professional vocal delivery, but also because they are Kenneth Branagh and Morgan Freeman. As such, their voices carry with them intertextual references to their previous roles and public identity. Prior to *Walking with Dinosaurs*, Branagh was better known for his roles in front of and behind the camera in various glossy Shakespearian adaptations. He also established his narrator credentials on both sides of the Atlantic by voicing the joint CNN–BBC television mini-series *Cold War* in 1998.[10] Through these associations, Branagh lends *Walking with Dinosaurs* the legitimacy and prestige gained through a long and successful career as a serious actor. Similarly, Freeman's oeuvre includes various roles as noble, trustworthy and likeable characters, and perhaps more importantly, as God in 2003's *Bruce Almighty* (Tom Shadyac), two years before he spoke the voice-of-God in *March of the Penguins*.[11]

More recently, documentaries with overt political agendas can be seen as capitalizing on their narrator's publicly expressed views and affiliations. *Inside Job*, Charles Ferguson's 2010 documentary about the 2008 global financial crisis, is narrated by Matt Damon, better known as the star of various Hollywood action films, including the *Bourne* franchise (2002–2016). In one interview with Ferguson, the interviewer suggests that he was reassured when he heard Damon's voice because 'he wouldn't take part in a documentary unless its research was solid and its assertions were true, so I can trust the statements in this narration'. Ferguson accedes this point, when he tells the interviewer 'it's not crazy that his voice gave you confidence in the film' and further implies that Damon was chosen to voice the film because he 'is known to be an intelligent man who has had some thoughtful involvement in politics' (Kors 2011). Of course what Ferguson does not acknowledge is that while Matt Damon has been outspoken about his left-wing political views, he is best known for sympathetic, heroic film roles. So, when he utters these words at the end of the film – 'the men in institutions that caused the crisis are still in power and that needs to change [...]. It won't be easy. But some things are worth fighting for' – perhaps they will resonate with audiences who subconsciously believe that that Matt Damon knows how to save the world. Thus, while this casting of Damon as the voice-of-God plays on the audience's intertextual knowledge of his real-life activities and political leaning, his on-screen persona looms large as well.

Conclusion

It is, I would suggest, becoming an increasingly popular tactic to cast a well-known voice as the voice-of-God. This tactic challenges the previously accepted evolution of documentary narration, one that was characterized by Charles Wolfe in 1997 as 'less assertive or homogenous than in documentaries of an earlier era: voices are personal or casual, multiple or split, fragmentary or self-interrogating, lacking a full knowledge of events or the motives and causal logic that a classical documentary would claim to disclose' (149). Instead, the famous voice-of-God remains as omniscient today as in the classical documentaries of the 1930s – no longer anonymous, but still revealing nothing of its speaker's inner self in the way that the fragmentary or self-interrogating voice-overs described by Wolfe do. There are, of course, commercial benefits to casting a well-known voice-of-God and it is a strategy that is predominantly used in mainstream documentaries that are bound either for primetime television slots or for theatrical releases.[12] If the aim for these documentaries is to get wider press coverage and higher box office returns or viewing figures, then the celebrity voice-over strategy seems often to be successful. *Inside Job* made $8 million worldwide against a reported $2 million budget – no mean feat for a documentary about economics. *March of the Penguins* made $127 million – the second highest grossing documentary of all time (after Michael Moore's *Fahrenheit 9/11* [2004]). For *The Eagle Huntress*, the addition of Ridley's narration was a strategic decision made prior to the film's release in order to widen its potential audience, and therefore box office returns. This suggests that star endorsement of a documentary in the form of contributing voice-of-God narration is a side effect of the emergence of the mainstream theatrically distributed documentary since the turn of the millennium, as these films are required to compete in a marketplace dominated by star-driven Hollywood fare.

However, the star documentary narrator also affords an opportunity to rethink the power of the voice-of-God, and indeed power more broadly. Rather than the authority coming from the anonymous and disembodied voice that abstractly represented the state in the 1930s classical documentaries, it now emanates from an identity that is recognizable from the voice alone. Star documentary narrators reside in some other space, one that is both off-screen and embodied. They remain in the imaginary wing, as Chion dubbed it, yet still the identity of the voice permeates the film, lending its position of social power and cultural capital to the documentary over which it presides. As such, the rise of the known documentary narrator indicates a shift in cultural power, from the state to the individual and is indicative, too, of the increasing rise of celebrity power in media and society more generally.

Notes

1 This chapter is focused on the English-language context.

2 Not all documentaries of the Griersonian era use the voice-of-God and images in the mutually reinforcing way that is seen in these two examples. For example, *The Song of Ceylon* (Basil Wright 1935) provides a subtle critique of colonialism through the counterpoint of soundtrack and image (Hoare 2010).

3 Aspects that were presented in *Byker* (Amber Collective 1983), a much later film that criticizes the regeneration of urban 'slum' terraces in Newcastle precisely because it destroyed the tightly knit local communities that were embedded in the areas so maligned in *To-Day We Live*.

4 These two accents are all but extinct now, with the evolution of both class politics and broadcasting standards meaning that a variety of accents are now heard on screen on both sides of the Atlantic. See British Library (N.D.) and Fallows (2015).

5 In fact, some of the imagery of the series is 'real' – in addition to the digitally created material, the series also shot in suitably 'prehistoric' – looking locations (such as New Zealand, the Bahamas and New Caledonia) and included the use of animatronic dinosaur puppets. This photographic material, however, is blended with the digital images so as to create a seamless, realistic whole and overall the series and the intertextual material surrounding it works hard to make the viewer

believe that if we could have filmed dinosaurs in the wild, this is what it would have looked like. See Honess Roe (2013).

6 In fact, we don't even know whether dinosaurs could vocalize. See Switek (2009).

7 Specifically, Memphis, Tennessee.

8 The universal adulation for Freeman's voice is indicated by the fact that he was chosen by popular public vote to voice Facebook CEO Mark Zuckerberg's private, in-home AI system. Lesser mortals can have Freeman guide them to their geographic destination via the GPS app Waze (Guynn 2016).

9 Although the subsequent popularity of the Attenborough version of the DVD suggests that such pandering was unnecessary. See Skipworth (2010).

10 Here, as in other significant ways, Branagh's career path has been observed as having much in common with that of Laurence Olivier. This is a comparison that is frequently, and often unfavourably, made in the popular press, cemented not least by Branagh playing Olivier in the 2011 film *My Week with Marilyn* (Simon Curtis). See Crompton (2017).

11 Freeman's association with the divine has been further perpetuated by his more recent appearance in the *National Geographic* documentary series *The Story of God with Morgan Freeman* (2016–present) in which he 'seeks to discover who God is'. http://channel.nationalgeographic.com/the-story-of-god-with-morgan-freeman/ (accessed 5 December 2017). Such intertextuality can be seen as coming full-circle in, for example, the use of Sigourney Weaver as the voice of the p.a. at the marine park in Pixar's *Finding Dory* (2016). This casting suggests that Weaver, through her previous voice work on *Planet Earth*, has become synonymous with the voice of natural history (for US audiences, at least).

12 Here, we could say that documentaries are taking a leaf out of animation's book, where well-known stars are routinely cast in voice roles following Disney's success with Robin Williams as the Genie in *Aladdin* (1993).

References

Banks-Smith, N. (1999), 'Roar of Approval: Last Night's TV', *Guardian*, 5 October. Available online: http://www.guardian.co.uk/media/1999/oct/05/tvandradio.television2 (accessed 5 December 2017).

Barthes, R. (1977), *Image, Music, Text*, London: Fontana.

Beattie, K. (2008), *Documentary Display: Re-Viewing Nonfiction Film and Video*, London: Wallflower Press.

Blair, O. (2016), 'Morgan Freeman Explains How He Got His Iconic Voice', *The Independent*, 4 March. Available online: http://www.independent.co.uk/news/

people/morgan-freeman-explains-how-he-explains-got-his-iconic-voice-a6911141.html (accessed 4 December 2017).

British Library (ND), 'Received Pronunciation', *Sounds Familiar?* Available online: http://www.bl.uk/learning/langlit/sounds/find-out-more/received-pronunciation/ (accessed 5 December 2017).

Bruzzi, S. (2006), *New Documentary*, 2nd edn, London: Routledge.

Chion, M. (1999), *The Voice in Cinema*, New York: Columbia University Press.

Chion, M. (2009), *Film: A Sound Art*, New York: Columbia University Press.

Chow, L. (2006), 'The Actor's Voice', *Cineaste*, 31 (4): 33–35.

Crompton, S. (2017), 'Why We Should Be Comparing Kenneth Branagh to Laurence Olivier', *Whatsonstage.com*, 7 April. Available online: http://www.whatsonstage.com/london-theatre/news/kenneth-branagh-olivier-awards-laurence_43316.html (accessed 5 December 2017).

Doane, M. (1980), 'The Voice in the Cinema: The Articulation of Body and Space', *Yale French Studies*, 60: 33–50.

Dolar, M. (2006), *A Voice and Nothing More*, Cambridge, MA: MIT Press.

DS (2005), 'March of the Penguins: Time Out Says', *Timeout.com*, 1 July. Available online: https://www.timeout.com/us/film/march-of-the-penguins-2005 (accessed 5 December 2017).

Fallows, J. (2015), 'That Weirdo Announcer Voice Accent: Where It Came from and Why It Went Away', *The Atlantic*, 7 June. Available online: https://www.theatlantic.com/national/archive/2015/06/that-weirdo-announcer-voice-accent-where-it-came-from-and-why-it-went-away/395141/ (accessed 5 December 2017).

Gwynn, J. (2016), 'Morgan Freeman to Voice Mark Zuckerberg's Jarvis', *USA Today*, 20 December. Avaialble online: https://www.usatoday.com/story/tech/news/2016/12/20/morgan-freeman-voice-mark-zuckerbergs-jarvis/95656216/ (accessed 5 December 2017).

Hoare, J. (2010), 'Analysis', Song of Ceylon, *Colonial Film Archive: Moving Images of the British Empire*. Available online: http://www.colonialfilm.org.uk/node/486 (accessed 5 December 2017).

Honess Roe, A. (2013), *Animated Documentary*, London: Palgrave.

Karpf, A. (2006), *The Human Voice: The Story of a Remarkable Talent* [e-book], London: Bloomsbury.

Kors, J. (2011), 'Inside Job's Charles Ferguson Wins Oscar, Airs Frustrations with Obama', *Huffington Post*, 27 February. Available online: https://www.huffingtonpost.com/joshua-kors/oscar-nominee-charles-fer_b_828735.html (accessed 5 December 2017).

Kreiman, J. and D. Sidtis (2011), *Foundations of Voice Studies: An Interdisciplinary Approach to Voice Production and Perception*, Oxford: Wiley-Blackwell.

Lemire, C. (2005), "March of the Penguins' Surprisingly Moving', *Today*, 20 June. Available online: https://www.today.com/popculture/march-penguins-surprisingly-moving-wbna8292753 (accessed 5 December 2017).

McCluskey, M. (2015), 'Listen to Scarlett Johansson Sexily Reading Bible Verses', *Time*, 28 October. Available online: http://time.com/4091276/scarlett-johansson-sexy-reading-bible-verses (accessed 5 December 2017).

Oaklander, M. (2016), 'Science Explains Why You Love Morgan Freeman's Voice', *Time*, 23 February. Available online: http://time.com/4233926/morgan-freeman-voice-waze-science (accessed 5 December 2017).

Renov, M. (2004), *The Subject of Documentary*, Minneapolis, MN: University of Minnesota Press.

Ruoff, J. (1992), 'Conventions of Sound in Documentary', in R. Altman (ed.), *Sound Theory and Practice*, 217–234, New York: Routledge.

Shingler, M. (2012), *Star Studies: A Critical Guide*, London: BFI.

Skipworth, H. (2010), 'Attenborough Victorious in the Battle of Narrators', *The Telegraph*, 9 June. Available online: http://www.telegraph.co.uk/culture/culturenews/7803409/Attenborough-victorious-in-the-battle-of-narrators.html (accessed 5 December 2017).

Sonnenschein, D. (2001), *Sound Design: The Expressive Power of Music, Voice, and Sound Effects in Cinema*, Studio City, CA: Michael Wiese Productions.

Switek, B. (2009), 'Did Dinosaurs Roar?' *Smithsonian.com*, 13 July. Available online: https://www.smithsonianmag.com/science-nature/did-dinosaurs-roar-48438337 (accessed 5 December 2017).

Wolfe, C. (1997), 'Historicising the "Voice of God": The Place of Vocal Narration in Classical Documentary', *Film History*, 9 (2): 149–167.

2

Auditing the Call Centre Voice: Accented Speech and Listening in Sonali Gulati's *Nalini by Day, Nancy by Night* (2005)

Pooja Rangan

What can we learn from call centres about the place of voice in documentary? Although these two sites – the offshore call centre and documentary's vocal conventions – may seem at first to have little to do with each other, I will suggest that their conjunction tells us much about documentary as a site of auditory interpellation. My analysis focuses on the accented speech of non-native English speakers and the techniques of making these voices intelligible, both in the context of the accent neutralization programmes that call centre trainees undergo and in terms of the use of documentary conventions to neutralize and render transparent, or conversely to mark and stigmatize the affective disturbances of accented speech. Call centre documentaries like *Nalini by Day, Nancy by Night* (dir. Sonali Gulati, 2005), in which I ground my analysis, offer a rich vehicle for this inquiry, since they dramatize the tensions inherent in and the potential of documentary as an 'accented' genre of cinema (Naficy 2001) at both a thematic and a formal level. The internal contradictions of this film prompt me to reframe its textual point of view, or what documentary scholars might call its 'voice', as an *audit*: a mode of speaking and organizing that both produces and is produced by unspoken norms of perceptual discipline or attention. I ask: What disappears or is neutralized at the level of the speaking voice and the listening ear when we sense such a textual voice in documentary, and what can the 'placelessness' of accented speech in the age of the call centre tell us about the stakes of this disappearing act?

My inquiry into the structuring absences of accented speech and listening pivots on a reassessment of 'voice' as the prevailing metaphor for a textual point of view in documentary. Following John Mowitt (2015) I call this reassessment an 'audit' of the place of voice in documentary: a form of review that acknowledges the persistent logocentrism of voice as a metaphor, and the ways in which this metaphor informs the way documentary films listen for, organize and distribute the materiality of vocal sounds, as well as how they invite us to listen to the world.

From voice to audit

In English, as Mowitt notes, there is no satisfactory equivalent to 'point of view' in the auditory domain; a 'point of audition' is not entirely satisfying or coherent since 'point' means little in the sonic register (2015: 4). Mowitt therefore proposes the term 'audit' in an effort to develop an analogue of 'gaze' in the auditory or sonic domain: one that would roughly designate 'that which exceeds and conditions hearing and organizes the field of the audible' (5–6). Mowitt describes audit as a 'hearing', or a mode of perception that has a primordial tie with aesthetics or what Jacques Rancière calls 'the distribution of the sensible' (Rancière 2006: 12). Mowitt further elaborates that this 'hearing' refers not so much to a 'perception or sensible event' but to a 'fold where perception turns back or over on itself, traversing the faculty of hearing with the angle, the posture of listening' (5). This complicated statement requires a bit of parsing: listening and hearing are often conflated, but hearing is generally considered a more passive mode of auditory reception, whereas listening is considered to involve 'making an effort to hear something' (Rice 2015: 99). Mowitt's definition of the audit as the fold where the posture of listening traverses the faculty of hearing can thus be understood as an effort to argue that (a) aesthetic forms introduce thoroughly ideological, if unconscious, perceptual distinctions, hierarchies or distributions into the conceptual domain of sound, and (b) these aesthetic forms, and the perceptual attunements that they cultivate in their audience, are both produced by and productive of the discursive conditions of audibility in any given conjuncture.

Documentary is one of these aesthetic forms. When we approach it from Mowitt's perspective, the popular metaphor of the 'voice of documentary' can be understood not just as a passive description of a film's textual point of view but as an articulation of how documentary's textual conventions are both produced by and actively productive of the audit. In the field of documentary, voice, rather than point of view or gaze, is the prevailing metaphor for a documentary film's particular world view or social perspective. The popularity of the voice metaphor owes in large part to an early and influential essay by Bill Nichols titled 'The Voice of Documentary', in which Nichols defines 'the voice of documentary' as 'that which conveys to us a sense of a text's social point of view, of how it is speaking to us and how it is organizing the materials it is presenting to us' (1983: 18). But even as he associates voice with speech, Nichols hastens to add: '"Voice" is not restricted to any one code or feature such as dialogue or spoken commentary. Voice is perhaps akin to that intangible, moiré-like pattern formed by the unique interaction of all a film's codes, and it applies to all modes of documentary' (18).

I have written elsewhere about the widespread and largely uncritical adoption of Nichols's metaphor – which stands simultaneously for a film's perspective, its mode of address and its textual organization, even as it invokes the varied narrative possibilities and political stakes of 'giving voice' to actual speaking subjects – by documentary scholars and practitioners over the past thirty years (Rangan 2017). Rather than reiterate these arguments, in which I redefine voice as an 'audibility', extending Gilles Deleuze's analysis of 'visibility' in Michel Foucault's writings (Deleuze 1988: 52), I would now like to offer some additional reflections on voice. I am interested in how voice, as a sonic metaphor for point of view, has defined documentary realities, by giving rise to conventions that not only 'construct perceptual conditions of hearing' but also 'highlight some features of reality and hide others' (Novak and Sakakeeny 2015: 1). I find it noteworthy that Nichols's metaphorical use of 'voice' simultaneously signals and disavows the documentary genre's literal emphasis on speech and innovation of vocal conventions (such as voice-over commentary, dialogue and the interview) as its distinguishing textual feature. Nichols's mention of the term 'codes' to refer to these conventions (indicative of the linguistic and post-structuralist bent of film studies in the 1980s)[1] refuses any naïve empiricism regarding documentary's social ethic of 'giving voice' to actual speaking subjects *even as* his use of voice to invoke that 'essential albeit invisible something' that coheres a film together seems symptomatic of the enduring Western metaphysics of presence that attaches to the voice in all of its conventional forms (Chow 2013: 40).

The same contradiction characterizes Nichols's recent return to his voice metaphor in *Speaking Truths with Film* (2016). Nichols proposes here that one of the major historical shifts in the voice of documentary since the genre's inception, reflective of technological as well as social changes, is the shift away from a voice that aims to *speak* (often from a unilateral and anonymous perspective, as characterized by intertitles in early silent films or the convention of 'Voice of God' commentary) to one that '*listens* with compassion' to 'histories from below' (82; my emphasis). The contemporary voice of documentary conveys an 'embodied perspective' to the viewer (82) that foregrounds elements of affect and emotional engagement (84), including inflections, gestures and behaviour (75). Intriguingly, even as Nichols associates documentary's evolving horizons with the potential for an encounter with an embodied diversity of affects and gestures, he again employs a vocal metaphor to describe that which these sensations ultimately convey: an imperative, interpellative address. He argues: 'Unexpectedly, someone calls out: "Hey, you!" [...] to be addressed by a film – to sense that a film seeks to engage and speak to us about the world we share – functions as a hallmark of documentary film' (74). To elaborate, a documentary film's 'Hey, you!' may be explicit, even didactic (the filmmaker speaks to us directly or through a proxy commentator) or it may be implicit or tentative (the editing or the choice of interview subjects conjures a voice or offers a proposal about the world). But Nichols is arguing that despite these differences in who is speaking and how we perceive their address, what is common to *all* documentary films is the *sense* we get of an interpellative 'Hey, you!' that seeks to 'engage and speak to us about the world we share', soliciting a form of attention that *listens for a voice*.

This is an intriguing formulation because it suggests that what arrests our attention in documentary, what makes us turn around and pay attention in response to its 'Hey, you!' is the implied presence of verbal address, the *sense* of a voice speaking to us, even as our sense of that voice is increasingly fleshed out, embodied and relational, as it were. I would like to offer a restatement of this paradox: the interpellative force of verbal address – whether actual or implied – makes a textual documentary voice palpable and demands a *listening* or a channelling of attention towards this voice, even as the vehicle or medium that imbues it with force disappears under it. When it is restated in this way, the disappearing materiality of voice in documentary resonates with what Mladen Dolar argues is the fate of voice in the Western metaphysical tradition, and that is to disappear. 'The voice', Dolar writes, 'is the instrument, the vehicle, the medium, and the meaning is the goal. This gives rise to a spontaneous opposition where voice appears as materiality as opposed to the ideality of meaning. The ideality of meaning can emerge only through the materiality of the means, but the means does not seem to contribute to meaning' (2006: 15). Voice, in this tradition, is conflated with meaning, and specifically, with *logos*: the referential content or message of speech. The embodied dimensions of voicing, including the phonic qualities of the voice in speech (accent, intonations, timbre, affectations) and non-linguistic, gestural or pre-phonic utterances (for instance: sighs, echolalia, affectations, laughter, hiccups, stammers) are considered potential obstacles to voice, and thus to the sovereignty of the subject (Weidman 2015: 233) – *except* in those cases where social and political norms imbue this matter with the capacity to disappear, marking it as non-matter or as a neutral, seemingly immaterial norm (20).

Could we not think of the metaphysical discourse of voice as an *audit*, or, that which exceeds and conditions hearing, and organizes the field of the audible in documentary, in the form of the voice metaphor? Once we do so, other questions emerge: as a sonic metaphor for point of view, how has voice (as the residue of a logocentric tradition) given rise to aesthetic and rhetorical conventions of voicing in documentary that in turn amplify some features of reality and muffle others? What can we learn from these vocal conventions about what counts as a voice and the world such a voice exhorts us to share, and conversely, about the sonic and gestural matter that counts but does not count? Or, to put it in terms of the distinction between hearing and listening: What ideologies about what counts as a voice do we passively receive and tacitly legitimate when we actively attend to a textual voice in documentary and respond to its hail? What invisible and imperceptible material supports – or conventional arrangements of embodied sounds and gestures – induce us to 'prick up our ears' and *sense* such a textual voice or perspective?

Sonali Gulati's *Nalini by Day, Nancy by Night* brings concrete detail to these abstract questions. Gulati's film is representative of a subgenre of recent call centre documentaries that focalize accent as an embodied material support that imbues speaking voices with ideological meaning even as it is obligated to disappear. At a thematic level, this film pinpoints the 'neutral accent' in which call centre agents are trained as an important site in which the social, lingual and geopolitical norms of the audit are consolidated under conditions of neoliberal global capital even as these norms go unheard. Although in principle and ethos, Gulati works *against* the prevailing norms of the audit, the vocal hierarchies and distinctions that result from her attempt to standardize accented speech and render it transparent illustrate its powerful yet imperceptible influence. I turn now to a discussion of these issues, with the following proposal: the voice of a documentary film ('that which conveys to us a sense of a text's social point of view, of how it is speaking to us and how it is organizing the materials it is presenting to us', as Nichols puts it) becomes discernable as an audit (i.e. as a means not only of speaking but of listening to the world and organizing an attentional economy) only when there are embodied obstacles to voicing present, whether explicitly or implicitly, as a disturbance or an interference.

The auditory patdown

Gulati offers the following anecdote to explain why she was instigated to make her film: having grown accustomed to hearing her name mangled by American telemarketers ('Sonali Gelato', 'Somalia Gelatin'), she was caught off guard when she received a telephone call from an agent selling Visa Platinum credit cards who pronounced her name 'as perfectly as people back home', even though the agent introduced herself as 'Nancy Smith'. Gulati would learn, by engaging her in conversation, that her given name was Nalini and that she was calling Gulati in Philadelphia from a call centre in New Delhi, where Gulati grew up.

Gulati's is one of several non-fiction films produced in the early 2000s that investigate the vocal norms in which agents in Indian call centres are trained, including *Diverted to Delhi* (dir. Greg Stitt, 2002), *Bombay Calling* (dir. Samir Mallal and Ben Addelman, 2006) and *John and Jane* (dir. Ashim Ahluwalia, 2005). These films share a fascination with the 'neutral accent' of call centre agents as a type of mask that disguises and displaces their class origins, and thus as a symptom of the homogenizing and reterritorializing forces of neoliberal, 'Anglobalizing' capital. Each of these films goes behind the scenes of the call centre industry in India to show us the face behind the voice, so to speak. Thus, the democratizing impulse of 'giving a voice' (or 'listening with empathy' to 'histories from below', to borrow Nichols's language) takes the form, in these films, of 'giving an image' to the faceless, nameless call centre agents who communicate using pseudonyms and pseudo-voices. In their cultural, economic, textual and spectatorial contexts, these films closely resemble the exilic and diasporic modes of filmmaking that Hamid Naficy dubs 'accented cinema' (2001): they are produced by diasporic filmmakers who have been displaced from their places of origin, under interstitial and transnational contexts of production, and at a textual level they invite a spectatorial engagement that is attentive to linguistic heterogeneity and the necessity of translation. These shared textual and contextual features prompt me to refer to these call centre films as a documentary subgenre even though they vary widely in production context and idiom.[2] For reasons of space, I have chosen to isolate and focus on Gulati's film as exemplary of this documentary subgenre.

Linguistically speaking, accent is a fuzzy term that is colloquially used to reference a specific 'way of speaking' (Lippi Green 1997: 42; 44). Shilpa Dave, author of *Indian Accents*, explains: 'An accent involves verbal intonations that stress particular syllables so the manner of speaking is just as significant as what is being said … Accent not only includes tonal qualities but also involves word choice, arrangement of words, and cultural expressions that are rooted in national (and regional) expressions of identity' (2013: 2). Every native speaker of English has an accent demarcated in geographic and social terms, but 'when a native speaker of a language other than English acquires English, accent is used to refer to the breakthrough influence of native language phonology into the target language' (Lippi Green 1997: 42–43). The notion of an accent is thus inherently comparative, appearing only in comparison to 'normal' or standard speech; empirically there is no such thing as an unaccented voice or a neutral accent. India has a high concentration of English speakers – a legacy of its colonial past – which has made it an attractive location for the business process operations (BPO) of multinational corporations. However, the phonological influence of regional dialects, or the 'Indian accent', presents a hindrance, one that the call industry attempts to minimize by stressing a 'neutral accent' that is at once 'global' and 'standard' English (Maitra 2017: 7).

In its inception in the late 1990s, the call centre industry in India operated under a 'cloak of secrecy' because American and multinational corporations feared customer backlash for outsourcing their BPO (Carillo Rowe et al. 2013: 9). Aimee Carillo Rowe, Sheena Malhotra and Kimberlee Pérez note: 'Agents were instructed to act as if they were stationed in the United States: they were required to take pseudonyms, speak with American accents, and deny their geographical and social locations as Indians' (Carillo Rowe et al. 2013: 9). Although the mandate of secrecy has since fallen, the practice of masking call centre operations still permeates the Indian call centre industry. This mandate, intended to minimize aural and cultural dissonance for native-English-speaking customers in the United States, the UK, Canada or Australia takes the form of rigorous voice training, accent neutralization and cultural competency programmes that can last anywhere from two to eight weeks (10–11). Accent neutralization has marked class implications. In a recent ethnography of call centres in New Delhi, A. Aneesh points out that many call centre trainees tend to be middle- and lower-middle-class migrants from smaller Indian towns who are trained to perform a globalized class identity by adopting Anglicized pseudonyms, 'switching off' local and regional linguistic habits and imitating common features of English speech that persist, to degrees, in the four aforementioned dominant accents (2015: 6–8, 62; Maitra 2017). Accent neutralization aims to scrub off the marks of an accent's place of origin so that the hearer cannot place it (Aneesh 2015: 4); thus, the neutral accent does not allude to a pre-existing reality so much as it performatively calls into being a certain placelessness, an auditory non-place (59; also see Augé 1995).

Gulati's anecdote about Nalini/Nancy offers two insights regarding the audial interpellations of the neutral accent. First, this anecdote attests to the capacity for misdirection at the heart of interpellation. Gulati fails to recognize herself in the content of the call, that is, in the call of capitalism, and does not act accordingly. But rather than simply rejecting the call by hanging up (the usual response of unwilling customers), Gulati concentrates on the vocal medium and not the message being conveyed by the telemarketer. Nalini/Nancy's Indian accent makes Gulati suddenly aware of the material support and embodiment of the agent's voice as an obstinate residue or remainder of accent neutralization (Dolar 2006: 20). Here, another powerful interpellation works in place of the failed one: the 'correct' pronunciation of Gulati's name by Nalini/Nancy triggers an auditory sensation of linguistic and geographic (un)belonging for the filmmaker, hailing her as a New Delhi-ite, even though Gulati had fashioned a new class identity for herself as an expatriate student in the United States. In that instance, Gulati is constituted as a member of an 'Anglobalized' and increasingly placeless linguistic community as a response to a call – an imaginary community that is constituted in and through the auditory interpellation of accented speech.

Gulati's anecdote also points to a second register of the audit/audition that the practice of accent neutralization attempts to anticipate, mediate and obviate, however unsuccessfully, by minimizing the traces of the body in the voice. I am referring to the 'sound hermeneutic' (Altman 1992: 252) activated by an acousmatic voice, especially one that is accented or somehow marked in auditory terms as 'other'. We might think of this hermeneutic manoeuver as an 'auditory patdown' or frisking in which the listener endeavours to visualize the embodied source or location of an unseen voice based on the audible evidence of its 'skin tones' (Chow 2014: 7–9) – evidence that only becomes palpable because of its departure from a perceptual norm that is transparent and thus inaudible. It is precisely this racialized presence of the body in the voice that the call centre industry aims to obfuscate or 'mask' by training agents in a 'neutral accent' that simultaneously masks the accented nature of listening by coding the listening ear of its ideal auditor as a 'colour neutral' norm. As a listener, Gulati departs from this colour-neutral norm, and this triggers a minor response to the aforementioned sound hermeneutic: Gulati's detection of Nalini's accent serves as a point of narrative intrigue that structures her quest to 'unmask' Nalini's placeless, acousmatic phone voice by localizing it in a 'real' face, body and story – that is, to give her back her voice by restoring her identity.

How should we think of this attempt to give voice to – or to lend an ear to – socially constructed sonic identities associated with voicelessness? Do call centre documentaries like Gulati's resist the audit, by unmasking the 'neutral accent' of call centre agents as well as the myth of neutral listening as twinned illusions or obfuscations? Or do the Anglobalizing mediations of the call centre haunt their own excavations of phonic and linguistic heterogeneity, leading them to mask their own accentedness – their own audit – even as they pursue the goal of auditory disillusionment? Gulati's film brings into focus what I see as a central site of tension within as well as the potential of documentary: that between the destabilizing diversity of vocal soundings opened up by the documentary encounter and the lingering metaphysical imperative that filters, hierarchizes and neutralizes this diversity in the name of 'giving voice'. This tension, or dialectic, is not new. As Trinh. T. Minh-ha and Brenda Longfellow once argued with respect to ethnographic documentary, the modernizing drive of 'giving voice' is at once a symptom and vehicle of the objectification of the world through linguistic colonization. Longfellow notes that ethnographic films from 1930 to 1970 tended to reiterate the political binarisms of colonialism at an aesthetic level, as a 'separation of image and voice [...] where voice is ascribed to the white European investigator and image to the bodies of colonial others' (2004: 338). This binary resulted in other discriminations or exclusions; to quote Trinh: 'Language as voice and music – grain, tone, inflections, pauses, silences, repetitions – goes underground. Instead people from remote parts of the world are made accessible through dubbing/subtitling, transformed into English-speaking elements and brought into conformity with a definite mentality' (1991: 60).

The call centre documentary illustrates the continuation and mutation of these dynamics in the era of neoliberal, Anglobalizing capital: an era in which *voice*, and not image, has become a vehicle of the ongoing objectification and fragmentation of the (post)colonial other. The phenomenon of accent is enabling in this context: it introduces a spectrum and hierarchy of cultural values in relation to the 'how' of voicing (grain, tone, inflections etc.) and its linguistic 'contents' where Trinh sees a dualistic gridlock. One scene in particular from Gulati's film dramatizes, in both content and form, the internalization of such a spectrum as a hierarchy – one in which a 'neutral accent' however chimerical, is deemed indispensable for social advancement. My reading of this scene channels Gulati's own attention to the contradictions between the film's message (its voice) and its vocal medium (its accent), as well as the economies of attention that emerge from their cleave. I offer the following proposal: we can only discern the *audit* of a documentary film in its accentedness by paying attention to the neutralizing masks and accentuating marks that shape our sense of its textual voice.

'Going nowhere': Masking and marking

The scene takes place midway through Gulati's film and features an excerpt from a failed job interview. We learn at the outset of the scene, from Gulati's voice-over commentary, that it is the candidate's fifth attempt at procuring a job as a call centre agent, and we learn at its conclusion that he has failed the interview yet again, because of his thick North Indian accent. The interviewer, a young Indian woman who speaks English fluently, but whose accent is more difficult to place, remarks to Gulati, who is behind the camera, that the candidate's pronunciation is 'nowhere'.

I would like to draw the reader's attention to two observations regarding this scene that have to do with the audiovisual representation of accented speech – or, to restate this chapter's conceptual refrain, the manner in which Gulati's film audits the job interview (itself an audition of sorts). The first, which will be readily apparent, concerns the way the voice of the candidate, a non-native English speaker, is heard and judged by himself and his interviewer as lacking. The second may be less apparent, even imperceptible: I am intrigued by how this structure of self-consciousness manifests in Gulati's film in the form of textual signs or conventions (voice-over commentary, subtitles, captions) that produce a hierarchy of accented voices based on their proximity to an unspoken yet palpable norm. To allow close attention to these audiovisual choices in relation to the word choices of the three subjects, I have transcribed the interview scene in detail. The interview takes place in English and is captioned in English, except for Gulati's off-screen voice-over, which is not captioned or subtitled, and the interviewer's final statement, which is subtitled rather than captioned:

Gulati (off-screen voice-over heard over street scenes of New Delhi shot at night – no captions or subtitles): It hasn't taken me long to realize that call centre jobs are indeed the most sought after jobs in urban middle-class India today. I've learned that there are three million people scrambling to land a prestigious job as a telemarketer. I even met with a structural engineer who's desperately trying to get a job at a call centre. This is his fifth attempt at a job interview.

Candidate (framed in medium shot – captioned throughout): Spectramind, I was up in fourth round, fourth round I was out.

Interviewer (off-screen – captioned throughout): What went wrong? What do you think must have gone wrong in Spectramind?

Candidate: I had a heavy regional accent in my English.

Interviewer (off-screen): What kind of a regional accent are you talking about?

Candidate: North Indian accent … Then I tried copying National Geographic's channel. Whatever they spoke, I tried it. I bought a tape recorder. I spoke, I heard what my voice looks like. I tried to improve on it. **Image cuts to medium profile shot of interviewer.** I have a friend. She's working at EXL.[3] She helps me a lot in it. **Image cuts back to candidate.**

Interviewer (off-screen): Okay … Which was the last movie that you saw?

Candidate: Last movie was *Face/Off*. Just 2 days before I saw it. I had seen it before, in between I saw it.

Interviewer (off-screen): Two days before where?

Candidate: On the TV, yesterday night I think it was …

Interviewer (off-screen): Can you describe it?

Candidate: Okay.

Interviewer (off-screen): Narrate the movie to me.

Candidate: In that movie there are two main characters, Nicholas Cage and John Travolta. Nicholas Ca … sorry John Travolta, he's a cop, and this Nicholas Cage, he's a criminal. What he does … he makes arrangements with his friends, captures the other fellow, takes him to the hospital, they change their faces. All the other assistants …

Interviewer (interrupts from off-screen): Change their faces?

Candidate: Yes.

Interviewer (off-screen): How do they do that?

Candidate: Some medical technology. He feeds them and changes it.

Interviewer (off-screen): Mm hmm.

Candidate: Then what happens after that, all his assistants, he kills them. So now there are only two people who know. Nicholas Cage and John Travolta, which is the real one?

Interviewer (off-screen): Right, thank you.

Interviewer (gets up, leaves the room): Welcome. **Image cuts back to interviewer.**

Interviewer (to an unseen interlocutor, possibly Gulati – subtitled as 'When he spoke, his tongue used to hit his teeth. He had a lisp and he was stammering in between'): *To* ... it was like, he had a lisp also, *jab woh bolta tha to* his tongue used to hit his teeth. **She demonstrates:** 'duh-duh' ... *jab main ... voh ... lisp aa jaata hai na* ... Then he was stammering also *beech beech mein*. Then he said, 'I don't want to be a follower, I want to be a leader,' *To* ... pronunciation was, like, nowhere. Next candidate.

This seemingly unremarkable scene from Gulati's film stands out to me because, in her attempt to listen empathetically to the candidate's predicament, Gulati also visually reinforces his voice as 'lacking': his accented speech requires captioning in English to ensure its intelligibility. In the judgement of his interviewer, the candidate's pronunciation is 'nowhere'. The implication is that the young man is a poor candidate for a call centre job because he has a 'nowhere' voice, a voice whose audible geographic specificity portends limited class mobility for the speaker. If he wants to 'go somewhere' in life, that is, if he desires economic and geographic mobility, however illusory or virtual that mobility might be, he must be able to scrub, mask or disguise his identity: not his face, as one might surmise from his rather ingenious plot summary of the film *Face/Off*, but the audible vocal markers that 'place' his voice in a location associated with underdevelopment. But this candidate's larynx, teeth, tongue and neural wiring are so stubbornly moulded by his inhabitation of a regional Indian dialect that his accented English utterances tether him to a local identity that he cannot escape or efface.

I am also intrigued by the presence in this scene of a formal hierarchy of audiovisual conventions that positions the three speakers (Gulati, the female call centre agent and the candidate) at various rungs along a class system of accents. Gulati's expository first-person voice-over, which is in direct and seemingly unmediated conversation with the audience, is at the apex of the hierarchy. Gulati's speech is not captioned, subtitled or anchored in an image, even though it is not unaccented. Her American-influenced lexical stresses and drawling vocal delivery intimate her expatriate status as a film student in Philadelphia (a piece of information Gulati volunteers early on in the film) even as its melodious up-down intonations, and rounded vowel sounds, evidence of a British English-medium education, respectively hint at her South Asian nationality and middle to upper-middle-class status.

In contrast, every word uttered by the job candidate is captioned, lassoing our gaze to the content of his words lest we be distracted by his 'erroneous' syntax and pronunciation (he rolls his Ds and Rs and strings words together), which indicate the audible influence of his native language (my guess is Punjabi) on his manner of speaking English. Visually, these captions resemble subtitles, which were first used in a documentary context in the 1970s by the ethnographic filmmakers Timothy Asch, John Marshall, and David and Judith MacDougall. As an alternative to dubbed voice-over for translating indigenous dialogue, subtitles offered a way of retaining the auditory qualities of the original speech, giving access thereby to the intellectual and emotional lives of subjects, and ceding the authority and interpretive monopoly of a voice-over to the synchronous voice of the ethnographic subject (MacDougall 1998: 7, 116). But the visualization of the job candidate's accented English as captions/subtitles has an effect contrary to Gulati's democratic aim of restoring this subject's sovereignty by 'giving him an image' – that is, by showing us the 'real' names and faces behind the anonymized call centre voice. As David MacDougall notes, subtitles can establish what is written as more definitive than what is said, producing the effect of a dialogue between the filmmaker-translator and the audience 'about' non-Western people even when the latter are seen and heard speaking in their own voices. The visualization of the oral as subtitles, he argues, can make audiences visually word-oriented and dependent, inducing a sense of passivity that detracts from a more active engagement with the sonorous grain and non-verbal expressive range of the speaker (1998: 174–175).

In Gulati's film, too, the candidate is transformed into an object/image in a dialogue between Gulati and the interviewer, even as he is metaphorically 'given a voice'. The interviewer's voice is less easy to place, and this may well be because she has been trained in the 'neutral call centre accent' that mimics a mixed set of features of British, American, Australian and Canadian English and is thus ironically 'placeless', even as she criticizes the candidate's voice as a 'nowhere' voice. Her words are captioned as well, but the motivation for this decision could be technical, since she is off-screen and un-miced for much of the exchange, and we cannot see her moving lips. When she finally appears on screen, the interviewer's words are subtitled rather than captioned since she switches between Hindi and English when addressing Gulati, relaxing the contrived norms of the interview and acknowledging Gulati's bilingualism as well as their membership in a shared linguistic class that does not include the job candidate: a class capable of 'masking' its origins. Even though we lose some of the content of what the interviewer says in translation, the suggestion is that her words do not require captioning, since the *sound* of her voice, unlike that of the candidate, approaches the aspirational linguistic norm of a neutral, placeless accent that can 'go anywhere'. That Gulati has achieved this global standard of cosmopolitanism by travelling to and living in the United States (a status that few call centre agents, if any, do; see A. Aneesh 2015: 64) is confirmed at the end of her trip to New Delhi, when she is offered a job as a voice and accent trainer at a call centre.

In conclusion, how should we think of these textual signs or audiovisual conventions: Gulati's off-screen commentary, the subtitled speech of the interviewer and the captioned speech of the candidate? Are these perfectly ordinary and unremarkable conventions – conventions whose ordinariness signifies the arrival of documentary as an 'accented cinema'? Or are they techniques of accent neutralization? Are they a cloak or disguise that simultaneously mask the accented basis of some voices and mark or accentuate the 'otherness' of others, thus enabling an unspoken yet palpable norm – a textual voice – to emerge? To me they are both. Their unremarkable and conventional quality might even be evidence of the successful camouflage of the accented relation between this textual voice and the listening ear that it hails ('Hey, you!'). But like the call centre voice, perhaps these conventions are an imperfect disguise – an unsuccessful attempt to neutralize the film's accent – that points to the capacity for error and misdirection built into every interpellation. Here I'm reminded of Gulati's accidental misinterpellation by Nancy Smith, by the 'correct' pronunciation of Gulati's name by Nalini/Nancy, which calls her attention to the medium of the voice rather than the meaning being conveyed. Just as Gulati fails to recognize herself in the call of capitalism but recognizes herself as a member of an increasingly 'placeless' linguistic community that is both nowhere and everywhere, I find myself similarly misinterpellated by the neutralizing masks and accentuating textual marks of Gulati's documentary, which call my attention not to the film's voice but to its *audit*. I'd like to think that confronting these conditions of hearing and the economies of attention that they sustain, or conversely, fail to contain, could replace 'voice' as a metaphor for how documentaries hail our attention.

Acknowledgements

I am immensely grateful to my colleague Amelie Hastie for her astute and eye-opening comments on multiple drafts of this chapter. I also thank audiences at

University of California, Berkeley, and University of Southern California, whose feedback helped me to realize this chapter in its final form.

Notes

1 I thank Michael Renov for this observation.
2 In contrast to the other films, which are fairly conventional documentaries shot on video, Ahluwalia's film is shot on 35mm and includes documentary material originally shot on video that was subsequently re-enacted and re-shot on film, producing an observational effect that is estranging and uncanny, much like the call centre voice itself. This interesting formal choice deserves a separate and longer discussion, but I will suggest here that the more conventional treatments of this voice, of which Gulati's film is exemplary, can be equally estranging. I hope to discuss these other examples of the call centre documentary subgenre in another venue.
3 The caption notes: EXL (call centre).

References

Altman. R. (1992), *Sound Theory, Sound Practice*, New York: Routledge.
Aneesh, A. (2015), *Neutral Accent: How Language, Labor and Life Became Global*, Durham, NC: Duke University Press.
Augé, M. (1995), *Non-Places: An Introduction to Supermodernity*, New York: Verso.
Carillo Rowe, A., S. Malhotra and K. Pérez (2013), *Answer the Call: Virtual Migration in Indian Call Centers*, Minneapolis, MN: University of Minnesota Press.
Chow, R. (2013), 'After the Passage of the Beast: "False Documentary" Aspirations, Acousmatic Complications', in P. Bowman (ed.), *Rancière and Film*, 34–52, Edinburg: Edinburgh University Press.
Chow, R. (2014), *Not Like a Native Speaker: On Languaging as a Postcolonial Experience*, New York: Columbia University Press.
Dave, S. (2013), *Indian Accents: Brown Voice and Racial Performance in American Television and Film*, Urbana-Champaign, IL: University of Illinois Press.
Deleuze, G. (1988), *Foucault*, Minneapolis, MN: University of Minnesota Press.
Dolar, M. (2006), *A Voice and Nothing More*, Cambridge, MA: MIT Press.
Lippi-Green, R. (1997), *English with an Accent: Language, Ideology, and Discrimination in the United States*, New York: Routledge.
Longfellow, B. (2004), 'The Great Dance: Translating the Foreign in Ethnographic Film', in A. Egoyan and I. Balfour (eds), *Subtitles: On the Foreignness of Film*, 335–353. Cambridge, MA: MIT Press.

MacDougall, D. (1998), *Transcultural Cinema*, Princeton, NJ: Princeton University Press.

Maitra, A. (2017), 'The Persistence of the Other: Revisiting the (Post-)Colonial Scene of Fragmentation' (Unpublished Conference Paper), 1–8, *American Comparative Literature Association Conference*, Utrecht, Netherlands, 6–9 July 2017.

Mowitt, J. (2015), *Sounds: The Ambient Humanities*, Oakland, CA: University of California Press.

Naficy, H. (2001), *An Accented Cinema: Exilic and Diasporic Filmmaking*, Princeton, NJ: Princeton University Press.

Nichols, B. (1983), 'The Voice of Documentary', *Film Quarterly*, 36 (6): 17–30.

Nichols, B. (2016), *Speaking Truths with Film: Evidence, Ethics, Politics in Documentary*, Oakland, CA: University of California Press.

Novak, D. and M. Sakakeeny (2015), 'Introduction', in D. Novak and M. Sakakeeny (eds), *Keywords in Sound*, 1–11, Durham, NC: Duke University Press.

Rancière, J. (2006), *The Politics of Aesthetics*, London: Bloomsbury Academic.

Rangan, P. (2017), 'Audibilities: Voice and Listening in the Penumbra of Documentary: An Introduction', in P. Rangan and G. Yue (eds), Special Issue on Documentary Audibilities, *Discourse* 39 (3): 279–291.

Rice, T. (2015), 'Listening', in D. Novak and M. Sakakeeny (eds), *Keywords in Sound*, 99–111, Durham, NC: Duke University Press.

Trinh, T. M. (1991), *When the Moon Waxes Red*, New York: Routledge.

Weidman, A. (2015), 'Voice', in D. Novak and M. Sakakeeny (eds), *Keywords in Sound*, 232–245, Durham, NC: Duke University Press.

3

The Fundamental Lie: Lip Sync, Dubbing, Ventriloquism and the Othering of Voice in Documentary Media

Patrik Sjöberg

This chapter will address the complex relationship between the represented body and face of a subject and the vocal expressions attached to them in documentary media. I will do this by examining instances in documentary where one voice has been substituted for another, or rather, where a voice has been attached to a body to which it doesn't belong. The discussion borrows from current research on the voice, the sonic arts, dubbing, ventriloquism, media history and lip sync. Although several works are mentioned in the discussion, this chapter focuses on a series of short films, *Covers* (2006 – ongoing), by artist and filmmaker Adie Russell. In her series, Russell re-enacts interviews and conversations with famous people from the past through her individual performance and lip sync in front of a camera. Russell's films lend themselves to a discussion of audiovisual representations of embodied speech acts and what Rick Altman (1980: 6) once referred to as 'the fundamental lie' at the heart of sound film; that is, 'the implication that sound is produced by the image when in fact it remains independent of it'.

In a brief description of her works on her website, Russell accentuates their 'temporal spaces'. These spaces get superimposed in the films, Russell suggests, because 'the historical moment of the audio recording' shares the space, both with 'the moment inhabited by myself and the viewer in the present'. Russell is well aware of the complexities that arise from such a seemingly simple trick of voice displacement. What is at stake, and what my discussion here elaborates on, includes questions of identity and performance; the encounter with archival audio; mediations of speech acts and the idea of an unstable audiovisual speaking subject. Russell's work not only challenges the idea that a single voice belongs to a single body, but also displaces gender and history, as all of the performed voices were recorded in the 1950s, 1960s and 1970s and the voices are all male. For Russell, the instability of the speaking subject is demonstrated through the unhinged voice, a voice unable to properly attach itself to either a historical or contemporary subject. Russell writes:

> The exchange between these moments exists as a composite of past and present, a kind of third space where meanings shift, judgements waver, visual cues are displaced, and the language hovers, *unattached to the identity of the original speaker, and yet not quite able to attach to my identity, my image, my mouth, my gestures.* (Russell, my emphasis)

In *Never Had More Trouble* (2007), Russell re-enacts a 1967 dialogue between Merv Griffin and Richard Nixon; in *The Reveal* (2012), she re-enacts a 1973 conversation between Dick Cavett and Marlon Brando; in *Time Can Do So Much* (2017), she (re)performs a conversation with Andy Kaufman; and in *The Most Extreme Perfect That Exists* (2011), she lip-syncs the words of Ingmar Bergman retelling an experience he had at a hospital.

An additional lip sync text this chapter explores is Gillian Wearing's *2 into 1* (1997). In Wearing's film a mother and her twin sons lip sync comments that one of the other parties has made about them in a prior recording when they were asked to speak candidly to the camera without the others present. The result is both moving and uncanny, so unmistakably intimate, yet distinctly reflexive and form conscious. In this and many of the works discussed here, this oscillation, this push–pull movement within the viewing experience makes us, on the one hand, acutely aware of the analytical and discursively informed strategy of the piece. And yet, on the other hand, we are drawn in, eager to engage with what these highly irregular speaking subjects have to say or perform.

How can we theorize this oscillation? Can these works inform us more broadly about speaking subjects in documentary media? What conceptual tools help us to analyse these films? In her preface to *Voice – Vocal Aesthetics in Digital Arts and Media*, Norie Neumark points out some central questions for this discussion:

> How do voices relate to bodies and subjectivity – how do voices speak to, from, and of embodied subjects? What happens to the embodied voice when it is mediated? How do voices speak of alterity – performing intersubjectivity, and sounding out the physical, affective, signifying and psychic spaces between subjects? (Neumark 2010: xvi)

My discussion of Russell's films in particular will offer ways of thinking about the questions raised by Neumark, in particular what historically and gender-displaced voices can teach us about the mediated, embodied voice. We tend to think of our voice as 'a voice' – a single entity – even though our voices are in a state of constant change: we age, we have colds, our accents are cut loose and set adrift to change depending on where we live; we smoke, we stay up late, we speak different languages, which significantly changes many things about our voice. We sound different when we are singing, crying or giving a presentation in front of an audience. We sound different in the morning than at night, we sound different recorded than live – and yet, we understand ourselves as having *a* voice, singular. However, contemporary discourse surrounding the voice, in work by Stephen Connor, Mladen Dolar, Norie Neumark and Adriana Cavarero, spans various disciplinary perspectives and yet seems to embrace the connection between voice, body, identity, speech acts and recordings as always already plural and contingent. In *Vocal Tracks: Performance and Sound Media*, Jacob Smith traces two strains in current discourse on voice; the first is distinctly focused on cinema and draws heavily on psychoanalytical theory to explore media objects' capacity for representing gender and identity; the second focuses on the cultural history of sound media and its implications for cultures of hearing. Smith identifies a central problem in both of these strands: the first tends to lose sight of the historical, social and cultural context of the media text, and the second tends to lose sight of the media object itself. Smith aims to avoid these issues by drawing insights from both of these strains and at the same time introducing theories of performance to the discussion. I align myself closely to Smith's position in my discussion of unhinged and swapped voices in documentary (Smith 2008: 4–5). Brandon Labelle, another prominent theorist of the voice, articulates the voice in relation to Lacanian theory: 'The history of the modern voice is the history of the split subject, the multiplication of consciousness, identity as always already divided, and by which voice is an operation that traverses these divides' (Labelle 2014: 90). The speaking subject we encounter in Russell's films can be viewed as such a split subject: the face and body of a young woman in the (relative) present performs a recorded voice of a much older man, now deceased, from the past in perfect lip sync and with a realist – that is to say, not exaggerated or obviously self-reflexive – rendering of facial expressions and body language.

The works by Russell in particular draw attention to the split between what we visually perceive to be the (single and permanent) host body of a voice and the voice we actually hear. Voices not only change, they can be exchanged.[1] To better understand the works analysed in this chapter, the relationship between voice and body needs to be placed in a media historical context.

Talkies, sound film conventions and lip sync

Apart from quality and volume, the challenge of the first sound film technologies was synchronization: the voice needed to fit the face, and fit it perfectly. The audience was by now used to faces without voices in silent cinema, and at the same time had heard recorded voices on the gramophone (and later, on the radio), so the notion that the voice was added on was prevalent. K.J. Donnelly discusses the way early sound film aimed for perfection in synchronization of voice and face and established the use of the so-called sync points in the film, which became an important formal element. According to Donnelly the 'spell of the diegetic (the "illusion")' becomes severed if the timing is off, but we as spectators can accept very slight delays by 'compensating' and focusing on other things (Donnelly 2014: 3–4).[2] When watching Russell's and Wearing's films, the synchronization is so close to perfect that the few instances when the timing is slightly off remind the spectator of the film's formal strategy of mismatched voice and body, and attention is drawn to the film's conceptual aims. Rather than follow exactly what the voices are saying, spectators are likely to engage in a version of the cinema of attractions (Gunning 2000: 229–236).

Apart from humorous pieces that play with the notion of poor dubbing (often found, for example, in 1970s Hong Kong kung-fu films), we rarely encounter deliberate asyncronicity between voice and body, and more specifically, voice and mouth. I am not considering the almost complete abandoning of all ties between body and voice, as found in Marguritte Duras's *India Song* (1975), but rather exploring cases that deliberately undermine conventions by falling in and out of lip sync. In his work on lip sync in punk rock films, David Laderman introduces a new way to look at live performance, authenticity and lip sync. A certain *sloppiness*, a more or less intentional loss of sync (or other sound-quality-related issues), brings out a certain otherness to the spectacle at hand, one that seems to both embrace and critique the spectacle. Laderman refers to this as '(s)lip-sync' where the performer deliberately shows that he or she is not singing live but performing a playback. For Laderman, the performer is 'undermining both the commodity fetishism and the underlying trope of authenticity intrinsic to conventional lip-sync', and he observes how these moments enact 'a split in the show, a rupture involving a double excess: the performer exceeds the performance spectacle by disengaging; the performance spectacle exceeds the performer by continuing – consuming, in a sense – the authenticity signified by the performer's dissociation' (Laderman 2008: 272). The few instances in Russell's film where she momentarily falls out of the otherwise almost flawless lip sync create a disruption in the already fragile and unconventional voice–face relationship. I consider Russell's slips closely related to (s)lip sync, as Laderman understands it. Whether intentional or not, these slips introduce a rift between face and voice that can become a meta-critical, and therefore memorable, moment for the viewer.

For Michel Chion, something happened to the body–voice relationship in the 1970s, when filmmakers gave up on the idea of trying to find *right* voice for the *right* body and instead came to the conclusion that 'there is no appropriate voice; every voice is a construction, a specific composition with the body' (Chion 2009: 142). Commenting on the drive to sync the voice to the body throughout film history (and thereby also partly contradicting Chion), Mark Kerins (2000) suggests that it was surround sound that enabled sync to be realized on a new level and that the synchronization of voice and body in cinematic works remains incomplete: the two may be *temporally* synchronized but, thanks to the surround sound technology, they may occupy different *spatial* positions. Kerins's argument can help us to further consider the voice–face relation in Russell's and other work that highlight the complexities of mediated voices, subjects, faces and their relation to technology.

Today, this gluing of voice to a face and body is standard practice across film genres and the close monitoring of the voices of characters in fiction film is handled by evermore sophisticated software. If the line delivered by an actor in a scene was not exactly what the director wished for, the actor no longer needs to visit a studio and read the line again and try to synchronize it with the lips' movement. Sound mixers can either lift the line from another take and simply make it fit the moving lips, or they can construct the phoneme they are looking for by using the archive of sounds that exist, drawn from all the other takes the actor has already recorded. The spoken word can be broken down to such small components that it can be moved around, tweaked and exchanged for others, and, most importantly, the lip sync kept intact.[3]

Voice and mediation

Neumark traces a connection between Kaja Silverman, Adriana Cavarero and Steven Shaviro in that they all embrace and elaborate on the way the voice can, and will, transgress the dichotomies of interior and exterior. Silverman emphasizes the way voice tends to be internal and external simultaneously, and thereby 'violating the bodily limits upon which classic subjectivity depends' (Silverman quoted in Neumark 2010: xix). This fluid and unstable state of the voice in its relation to a body is also emphasized by Steven Shaviro, who writes that the 'voice always stands in between: in between body and language, in between biology and culture, in between inside and outside, in between subject and other, in between mere sound or noise and meaningful articulation' (Shaviro quoted in Neumark 2010: xx). These thoughts are complemented in Neumark's text by Cavarero's challenge to Jacques Derrida's understanding of presence by instead articulating an alterity connected to her understanding of intersubjectivity (Neumark 2010: xix–xx).

Perhaps it is in this in-betweenness that we can locate the point of departure for the type of voice–body relationships evident in Russell's films, where unhinged voices haunt alien bodies. Neumark identifies a process she calls 'unvoicing' – understood as 'playing between presence and absence' where 'there is something else, something more, something other, which unvoice performatively brings into play' (Neumark 2017: 124). What is present, then, in Russell's films, is Russell's body and also her unvoice, just as there is a presence of the voice of the recording she lip-syncs, with its own 'voice-body' as so-named by Steven Connor. Lining up the voice of the deceased subject, present through the archival recording, with the living body at the time of the film's recording highlights not only the way the body and voice are present in different capacities, but also the acute absence of the missing voice of the visible subject and the missing body of the archival voice.

In *The New Soundscape*, Murray Schafer (1969) coins the term 'schizophonia' to mean the cutting free of a sound from its natural origin; this term has achieved a certain measure of success in naming instances in which sounds, including the voice, are dissociated from their sources. Schizophonia, however, should not be viewed as synonymous with Chion's influential term *accousmêtre*, which refers to the disembodied voice. Chion's term is less helpful for the discussion here, as the notion of the disembodied voice does not exist in documentary work in the same way as in fiction, which is where Chion situates this term (see Honess Roe in this volume). In place of the disembodied voice, I prefer the understanding of *the body implied* as expressed by Connor. Commenting on his own influential book, *Dumbstruck–The Cultural History of Ventriloquism*, Connor remarks:

> Not only had I shown how much violence there often is in the dissociation of the voice from the body, I also thought I had at the end of *Dumbstruck* earned the conclusion that there are no disembodied voices; for every disembodied voice is always also what I called a 'voice-body', the body implied by or intuited from the voice. Every voice, it seemed to me, was a kind of auditory physiognomy, not in the usual sense of what the body makes known through outer indications, but in the sense of the making known of a body that is undertaken by every voice. (Connor 2012)

Here Connor articulates something that expands, or perhaps questions, the notion of the acousmatic: the implied body indicates a phantasmatic host or an origin for the voice. We imagine, or construct, a ghostlike body to which we attach the voice. In Russell's films, the original body belonging to the recorded voice has been swapped for another, clearly mismatched body: that of a woman who performs the recorded voices of men like Richard Nixon, Marlon Brando, Andy Kaufman and Ingmar Bergman. Although the pieces are not very long, ranging between two and twenty minutes, the relationship between the body and face of the young woman lip-syncing the voice of the older man (recorded several decades earlier) changes as the films progress. Once the initial fascination with the near-perfect lip-syncing, and the strange, funny and uncanny combination of voice and body has subsided, we accept the speaking subject and start to listen to what is being said instead of focusing on the performance. Without ever completely forgetting the strategy deployed by Russell, we accept this voice–body relation and its ability to speak to us from documentary rather than fiction. The speaking subject is not Russell, nor Brando, but a third. The ability to accept, or construct, this particular position might partly rely on the operation Connor alludes to when he talks about the way we construct the 'voice-body', or the body implied – and yet it moves beyond Connor's concept. The implied, or performed, voice-bodies we meet in Russell's work give us the opportunity to consider this process in works that do not engage these issues quite so concretely; I'm here considering the global cultures of dubbing or voice matching in animation, for example.

Vocal displacement in documentary

Two films that have been noted for their intricate and deliberate use of voice-play and lip sync to address the voice–body relationship in documentary are Clio Barnard's *The Arbor* (2010) and Peter Middleton and James Spiney's *Notes on Blindness* (2016). *The Arbor* fuses documentary with fiction to tell the tragic story of playwright Andrea Dunbar, who died at the age of twenty-nine. The documentary subjects are played by actors, but the voices heard belong to the actual subjects and were pre-recorded in interviews and the actors lip-sync their words on screen. A similar strategy can be found in *Notes on Blindness*, but the latter film has a different appearance and style: whereas *The Arbor* never departs from its reflexive presentation, *Notes on Blindness* allows the spectator, at least at times, to follow the lip-synced characters in a more classical mode of transparent narration where the mode of production is not highlighted but hidden. While both *The Arbor* and *Notes on Blindness* rely on the close-to-perfection of the lip sync to construct their respective subjects, *The Arbor* uses a wide range of reflexive tools throughout the film in contrast to *Notes on Blindness*, which establishes a clearer diegetic world in which the subjects perform. Both films contrast with the highly reflexive work of Russell and Wearing, where there is no attempt towards realism in terms of surroundings or context to heighten a sense of illusion.

Both *The Arbor* and *Notes on Blindness* offer abstraction as a hidden potential of documentary conventions. In *The Arbor* and *Notes on Blindness*, just as in Russell's films, the strategy of matching a different body to the voice offers the spectator the opportunity to consider the face and body of the actor, on the one hand, and the voice, on the other hand, as both separate and unified entities. Or, more precisely, they are never entirely separate and not entirely one. This phenomenon has been noted in fiction film, particularly in *Singing in the Rain* (1952) and *The Exorcist* (1973), as well as in singing numbers in musicals more generally,[4] but the discussion is not as obvious for documentary. The *logocentrism* of documentary as a rhetorical form demands that we look more closely at films that challenge the conventional voice–body relation.[5] Documentary has often taken on the responsibility of speaking for others: representing marginal subjects in a fight for justice, for example. This has been understood as a problematic version of political ventriloquism. On a formal level, however, ventriloquism offers another way of thinking about the voice–body relationship.

Ventriloquism

Using ventriloquism as a discursive tool, or metaphor, is now well established across disciplines such as political science, anthropology and linguistics. David Goldblatt calls ventriloquism 'illusion without deception' and suggests that ventriloquism should be seen as a strategy for the construction of virtual possible selves (Goldblatt 2006: 176). Connor defines 'ventriloquism' as not merely making one's voice appear from somewhere else, but as a term that designates all dissociated or displaced voices, along with the various explications of such voices. In short, 'ventriloquism' is a general term for any variety of speaking for, or through, a designated and represented Other. A discursive model of ventriloquism found in work of these theorists is helpful for examining another aspect of the body–voice relation, namely the *ecstatic voice*.[6]

The strongest taxonomic link between these various 'ventriloquisms' is the signification of another identity to which the responsibility for speech is deflected. Charles Bruce Davis (1997) identifies a set of questions that are valid for the works discussed here. Davis focuses on questions relating to the act of speaking, such as 'Who did who think was speaking, and to whom?' This is a question of signification and reception, of identification in cultural and historical context, and of accountability for what acts of speech can do. Ventriloquism, Davis argues, foregrounds 'voice' not as individual expression or linguistic point of view, but as a signification of an identity that is always under construction in a give and take dialogue, not only with other voices but with every variety of cultural restraint which language and bodies place on communication.

Looking more closely at how Russell's and Wearing's work can be viewed through this ventriloqual perspective, we can begin with Connor's suggestion that there are three types of ventriloquisms: 'engastimyth', an ancient way of saying 'speaking from the belly' where the subject 'is demonically given or gives herself the voice of the other'; the modern ventriloquist, who 'gives his own voice to another'; and a mediating third, where the voice again is given to another 'but now it is the unwelcome and unsolicited gift of *her own* voice' (Connor 2000: 198). Russell's filmed voice performances would qualify as the archaic engastimyth. Here an unseen subject has provided the filmed body with a voice. In Wearing's *2 into 1* it is less clear: the bodies of the twin sons are present in the work, as is the body of the mother, as well as their respective voices; they have just been exchanged. The voice of another is exchanged for that of the visible subject, but the bodily origin of the voice also appears in the work. The connections between the respective bodies and voices remain intact by cutting between the performances. Neumark's question about the subjects in Timothy Vernon Moore's *The Forgotten Man* (2006) rings true for all the works mentioned: 'What has become of these characters, now ventriloquized, haunted, *inhabited* by TV Moore – or is it he *inhabited* by them?' (Neumark 2017: 134) Neumark echoes the questions posed by Russell: Where does the voice reside, to what body is the voice attached, and what happens to it if it is somehow unattached to a specific body that occupies an easily identified time and place?

Enacting history and identity through (re)performance

As Russell states in the introduction to her series of films, the historical character of the voices she is lip-syncing is central to her project. The discrepancies between the image of Russell and the words we hear work on several levels simultaneously: first we have the material differences in that we can hear that the audio was recorded decades before the visual image. With the footage from one era and the audio from another, each film's temporality seems to belong to neither past nor present. The audio is anchored in its time with a format and manner of speaking typical of mid-twentieth-century American talk shows. The footage is well anchored in an era closer to the time this is written. This is true of *Time Can Do So Much* (2017) in particular, as it was filmed on HD video, and the detailed crispness of the image and rendition of colour stand in sharp contrast to the obviously archival dialogue. The temporal indeterminacy of the speaking subject should be added to the temporality already inscribed in the voice itself, or as Brandon LaBelle understands it:

> The voice puts us on unsteady ground – for example, I'm speaking now, my voice has a presence, and at the same time it so easily dissipates, is forgotten and trails into silence. The voice only allows me to have presence by eternally leaving me behind. It is a kind of pre-echo of the body to come – it precedes me and at the same time is cut off from my body. (LaBelle 2010: 128–129)

The second layering has to do with the play of identity the works engage in as the younger woman lip-syncs an older man's voice, highlighting gender and age as relevant to the construction of a speaking subject.

Figure 3.1 Adie Russell as Merv Griffin. *Never Had More Trouble* (Adie Russell 2007). Courtesy Adie Russell.

Figure 3.2 Adie Rusell as Richard Nixon. *Never Had More Trouble* (Adie Russell 2007). Courtesy Adie Russell.

What makes both the temporal and gendered aspects of the film significant has everything to do with the way Russell performs the pre-recorded audio. She uses her wide range of facial expressions and her body language, timing the sighs, the coughing, the extended pauses and, not least, the listening pose of the other person in the interview. The better her performance is, the more directly she embodies the recorded voice, and thereby, comes closer to completing the illusion. Yet, such completion is never possible due to the temporal discrepancy between image and sound and the play with gender. In other words, she is performing in such a manner as if she indeed were trying to fool the spectator; at the same time, she knows, of course, that this is impossible. It is, in Goldblatt's words, *illusion without deception*. Nowhere is this clearer than when the speaking subject falters, pauses or looks for a word or the proper phrase. In the encounter between Richard Nixon and Merv Griffin, for example, the conversation is often tense and uncomfortable, particularly when Griffin suggests that Nixon is perceived as a loser. Here Russell uses hostile and defensive facial expressions that suggest she was actually saying the words. In the case with Ingmar Bergman's voice in *The Most Perfect Thing*, Russell has a distinctly Swedish accent to deal with, as well as the obvious trouble Bergman has finding the correct words in English. His habits of stuttering, slipping into Swedish and accepting words from a voice nearby are all meticulously acted out with perfect lip sync. The act of speaking is brought to the fore with some force in these scenes: the difficulties of speaking in a foreign language, bringing attention to the face and body as the struggle to find words is manifested in the physical act of speaking. In the conversation between Marlon Brando and Dick Cavett, played out in *The Reveal*, discomfort and low-intensity hostility are apparent almost immediately and last throughout most of the conversation. The distinct discrepancies in voice between Brando and Cavett are signalled only by a change of dress and hairstyle in Russell's appearance; she plays unmistakably different characters but her facial expression and body language remain the same throughout the performance. Russell's toned-down but naturalistic acting style produces two distinctly separate speaking subjects. It is almost a textbook example of how voices and bodies can be unpacked and reassembled with dramatic results.

In both *Time Can Do So Much* and *The Reveal*, there is also another, more thematic tangent to the discussion of authenticity, presence, performance and voice. In *Time Can Do So Much* we hear Kaufman talk of his fascination with Elvis Presley, whom he has been impersonating since he was a child. As an adult, Kaufman performed his version of Presley on television, and Russell's film ends by showing some of this footage.

Figure 3.3 Adie Russell as Andy Kaufmann. *Time Can Do So Much* (Adie Russell 2017). Courtesy Adie Russell.

The footage is slowed down and the voice we hear singing is the actual Presley, not Kaufman. Since the footage is shown in slow motion, there is no chance for proper lip sync to occur, which forms a sharp contrast to the preceding footage of Russell lip-syncing Kaufman's words. This adds an extra dimension to the speaking subject(s) we encounter in the work as a whole. In *The Reveal*, the conversation originates from a ventrilocal, not a lip-syncing, event. Awarded an Oscar in 1973 for his performance in *The Godfather*, Brando sent Sacheen Littlefeather to speak in his place on behalf of the historical mistreatment of Native Americans, including a critique of their representation on film. In both Kaufman's impersonation of Elvis, which alludes to the performance tradition of lip-syncing as well as impersonation and imitations, and in Brando's insistence on letting Littlefeather speak for him, we are reminded of Connor's understanding of ventriloquism as any variety of speaking for, or through, a designated and represented Other.

Towards a conclusion

Before any transplant of an organ from one person to another can be surgically executed, the doctors need to be sure that the host body will not reject the new organ. The match of blood types is only one of several things that need to align.

The matching of voices to bodies and faces in film cultures devoted to dubbing fiction film, or in musicals for that matter, seems to work according to a similar logic: age, dialects, intonation and sociolect are all there to facilitate the success of the transplantation.

In the works discussed here, and addressed within a context of documentary studies, the operating logic seems to be more *Frankensteinian*. Doctor Frankenstein, as we know, took parts from all kinds of bodies without paying too much attention to attributes or characteristics of the discrete parts; stitching them together and electrocuting the product made it come alive, regardless. Staying with this metaphor, one could argue that the bond between mother and children in Wearing's *2 into 1* could form one such matching bond, through blood and familiarity, but we are still left with a potential mismatch in the film involving age and gender, not to mention level of education and ability to articulate one's thoughts. In the end, Wearing's work may be more Freudian than Frankensteinian, in that it primarily elaborates on the mother–child relation rather than the meta-representational.

Russell's work, however, evidences the Frankensteinian impulse. The complicated speaking subject created by this particular construction of voice and body seemingly accepts transplantation from anyone; that is, the body will not reject any voice sutured onto it, and, at the same time, the voice will not reject any body stitched to it. The process seems to parallel Chion's term *synchresis*, where a forging bond is erected between the visual and the audio event regardless of its origin. Through the constructed performance of bodily and facial expression, and the carefully articulated lip sync of the archival voice from the past, mismatch of gender, identity and history are overcome to make the operation successful forming a new, densely layered speaking subject.

The Kuleshov effect holds that the meaning of an image of a facial expression does not reveal itself until it is sequenced with a second image: the meaning alters depending on the combination of images. Perhaps this is the most common basis for how we could consider the relationship between voice and implied bodies, even though the voice and implied body combination generally occurs as a simultaneous rather than a sequenced one. An otherwise hidden meaning of a statement or voice will be revealed when it is (mis-)matched with another body and face, just as the 'speaking' body will reveal other aspects of itself when paired with the voice of someone else. Also, as we see in the works of Russell and Wearing, a dislocation of voices, a mismatch of voices and bodies, in no way disqualifies the speaking subject or the position it speaks from. They do, on the other hand, invite us to consider the relationship between voice and face, voice and body, voice and subjectivity and the mediation of these relationships in documentary and how they are always open to creative recombination.

Notes

1 I also view this discussion of the documentaries in question as closely related to other instances in media history found outside the domain of film and documentary studies. For example, the development of the prosthetic larynx; the software and cultural analysis surrounding the electronically generated voice of Stephen Hawkins; the media reactions to the scandalous affair concerning the performers/imposters in Milli Vanilli (leading to the formation of The Real Milli Vanilli), to the media voice ban implemented by Margaret Thatcher on BBC regarding Gerry Adams between 1988 and 1994. They all, in different ways and in different media outlets, as well as in different political, social and cultural contexts, contribute to a discussion of authenticity, identity and voice displacement.

2 In countries where the dubbing of voices is the norm, this slight mismatch, or loss of timing, is very much part of the cinematic experience.

3 For an overview see Sonnenshein (2001).

4 This discussion is successfully developed by Jennifer Fleeger (2014). We are in fact inclined to accept considerable dissimilarities between, for example, the spoken voice and the singing voice in many musical films, perhaps, partly, because there is a range of aspects of the film that we also have to accept to make this genre work. It is still interesting to consider the way the voice of Marni Nixon, for example, was used for the singing instead of Deborah Kerr in *The King and I* (Walter Lang 1956), for Nathalie Wood in *West Side Story* (Jerome Robbins and Robert Wise 1961), Audrey Hepburn in *My Fair Lady* (George Cukor 1964) and Marilyn Monroe in *Gentlemen Prefer Blondes* (Howard Hawks 1953). Attesting to our ability, or desire, to attach voices to bodies, the singing voice of India Adams was lip-synced by both Cyd Charisse in *The Band Wagon* (Vincente Minelli 1953) and Joan Crawford in *Torch Song* (Charles Walters 1953), in the same year, 1953, without drawing attention to it.

5 An overview and commentary of this trajectory in documentary studies can be found in Rangan (2017: 134–141).

6 The notion of ecstatic voice comes from the Greek *ekstasis*, which translates to 'standing apart from oneself'. The term 'ecstatic voice' is most commonly used in religious and anthropological contexts and relates to situations where a person is, so-called, speaking in tongues or in other ways vocally channelling someone else, often some form of supernatural entity or divinity.

References

Altman, R. (1980), 'Moving Lips: Cinema as Ventriloquism', *Yale French Studies* 60: 67–79

Chion, M. (1999), *The Voice in Cinema*, New York: Columbia University Press.

Chion, M. (2009), *Film: A Sound Art*, trans. C. Gorbman, New York: Columbia University Press.

Connor, S. (2000), *Dumbstruck – A Cultural History of Ventriloquism*, Oxford: Oxford University Press.

Connor, S. (2012), 'Panophonia', talk given at the Pompidou Centre, 22 February. Available online: http://stevenconnor.com/panophonia.html (accessed 17 October 2017).

Davis, C. B. (1997), *Ventriloquism: Identity and the Multiple Voice*, PhD diss., University of Washington.

Dolar, M. (2006), *A Voice and Nothing More*, Cambridge, MA: MIT Press.

Donnelly, K. J. (2014), *Occult Aesthetics: Synchronization in Sound Film*, Oxford: Oxford University Press.

Fleeger, J. (2014), *Mismatched Women – The Siren's Song through the Machine*, Oxford: Oxford University Press

Goldblatt, D. (2006), *Art and Ventriloquism*, London: Routledge.

Gunning, T. (2000), 'The Cinema of Attraction: Early Film, Its Spectator, and the Avant-Garde', in R. Stam and T. Miller (eds), *Film and Theory: An Anthology*, 229–236, Oxford: Blackwell.

Kerins, M. (2000), *Beyond Dolby (Stereo): Cinema in the Digital Sound Age*, Bloomington, IN: Indiana University Press.

LaBelle, B. (2010), 'Untying the Tongue, the Promise of the Voice from Vibresppasce to Public Space', in D. Diederichsen and C. Ruhm (eds), *Utopia of Sound: Immediacy and Non-Simultaneity*, 128–139, Vienna: Schlebrügge.

LaBelle, B. (2014), *Lexicon of the Mouth: Poetics and Politics of Voice and the Oral Imaginary*, New York: Bloomsbury.

Laderman, D. (2008), '(S)lip-Sync – Punk Rock Narrative Film and Post Modern Musical Performance', in J. Beck and T. Grajeda (eds), *Lowering the Boom: Critical Studies in Film Sound*, 269–288, Chicago, IL: University of Illinois Press.

Neumark, N. (2017), *Voicetracks: Attuning to Voice in Media and the Arts*, Cambridge, MA: MIT Press.

Neumark, N., R. Gibson and T. Van Leeuwen, eds (2010), *Voice: Vocal Aesthetics in Digital Arts and Media*, Cambridge, MA: MIT Press.

Rangan, P. (2017), *Immediations: The Humanitarian Impulse in Documentary*, Durham, NC: Duke University Press.

Russell, A., www.adierussell.com (accessed 20 October 2017).

Schafer, M. R. (1969), *The New Soundscape*, Scarborough, ON: Berandol Music Limited.

Smith, J. (2008), *Vocal Tracks: Performance and Sound Media*, Berkeley, CA: University of California Press.

Sonnenschein, D. (2001), *Sound Design: The Expressive Power of Music, Voice and Sound Effects in Cinema*, San Francisco, CA: Michael Wiese Productions.

Whittaker, T. and S. Wright, eds (2017), *Locating the Voice in Film: Critical Approaches and Global Practices*, Oxford: Oxford University Press.

Section Two

Voices beyond Language

4

Hearing Voices in Singer-Songwriter Music Documentaries

Maria Pramaggiore

Amid the growing popularity of documentary in the twenty-first century, the feature-length music documentary has been thriving. A spate of films released in the early 2000s sketches portraits of a diverse range of musicians including Janis Joplin, Nina Simone, Kurt Cobain, Bill Withers, Lady Gaga, James Brown, Ginger Baker, Glen Campbell, George Harrison, Amy Winehouse, Whitney Houston, Roky Erickson, George Michael, Daniel Johnston and Kathleen Hanna, and bands from the Sex Pistols, Metallica, Anvil and Journey to the Stooges. Music documentaries are regularly screened in theatres and at major film festivals; in early 2018, *Billboard* reported that 'major music artists will be driving much of the conversation' at Sundance (Dollar 2018). Although the theatrical and festival screenings of these projects challenges Kevin Donnelly's assertion that music documentaries are no longer viewed on the big screen (2015: 71), the proliferation of small-screen streaming providers, such as Netflix, HBO and Amazon, has certainly contributed to the genre's renaissance.

In this re-energized era for the music documentary the singer-songwriter portrait has asserted its dominance. Music critic Phil Gallo observed that 'back-to-back Oscar wins by […] 2012's *Searching for Sugar Man* and 2013's *Twenty Feet From Stardom* encouraged filmmakers to dive into passion projects about lesser-known musicians, obscured chapters of music history and the private lives of very public people' (2015: 121). Eschewing the audience-centred approach pioneered by D.A. Pennebaker's *Monterey Pop* (1968) and Michael Wadleigh's *Woodstock* (1970), films that attempt to recreate the psychedelic sensorium of the live performance for the film audience, these recent features, as Gallo's comment suggests, focus on the turbulent lives of musical performers, often deploying an intimate, narrative style. Borrowing from the Hollywood backstage musical, this new wave of portrait films adapts the televisual aesthetic of VH1's popular *Behind the Music* series, which aired more than 240 episodes between 1997 and 2017, and also draws inspiration from reality TV, which afforded washed up 1970s rockers Ozzie Osbourne and Gene Simmons second careers.

Devoting copious amounts of screen time to videos supplied by family, friends and fans, previously unreleased tracks and bootleg recordings, personal mix tapes, journals and drawings, these decidedly digital-era documentaries aspire to be completist portraits of the artist as singer-songwriter. They thus investigate not only the rhetorical voice of the songwriter as well as the performing/singing voice, but also encompass everyday, non-performing voice(s) that emerge from interviews, informal discussions and spontaneous home videos. The fact that these documentaries are as invested in the melodrama as they are in the music might imply that the latter is given short shrift. That's not the case, however; although they tend to favour a conventional narrative structure, hewing closely to a familiar 'A Star is Born' saga of rise and fall, the films nevertheless grant primacy to writing and performing music, explicitly linking music to performers' private, 'authentic' identities as well as to their professional triumphs and tribulations. These singer-songwriter documentaries resoundingly reject 'the perennial notion of sound being secondary in the storytelling process' (Beck with Ament 2015: 117), because observing the activities associated with making music is understood as offering access to the performer's subjective experience.

As examples of 'character driven documentary' (Cagle 2012: 54), these films often identify the source of musical creativity in the pain of difficult childhood experiences or social alienation and invariably document the challenges of coping with celebrity. Drawing from a seemingly endless archive of performances, interviews, home movies and fan videos, these documentaries immerse viewers in the singing, songwriting, interviewed and everyday voices of the performers they depict. In sensory terms, this crowd-sourced sonic aesthetic probably falls short of the immersive experiences touted by Virtual Reality and other new media technologies, 'whereby the viewer becomes totally enveloped within and transformed by "the virtual environment"' (Dyson 2009: 1). Yet the attention these documentaries pay to the attributes of the human voice bespeaks an attempt to create, document and immerse viewers in disembodied, extrasubjective sonic encounters as a means of acquiring a deeper understanding of the performer.

This chapter argues that singer-songwriter music documentaries employ strategies of vocal excess and immersion in order to appeal to fans' desires to gain privileged access to, or indeed embed themselves within, performing subjectivities. Ironically, when they go behind the scenes to listen to the voices of musicians offstage, these films reveal that performing voices are complex temporal and spatial entities, not singular, fixed, definable objects. As much as fans seek out films that can reveal the mysteries explaining their favourite musicians' creative gift in the hopes of apprehending a 'true self', the sonic attributes of voices and selected techniques of sound design can resist that process. The observations that Frances Dyson makes about 'a' sound are also true of 'a' voice: '"A" sound is always multiple, heterogeneous; being neither visible nor tangible, sound is never quite an object, never a full guarantor of knowledge' (2009: 4–5). The temporal ephemerality and spatial diffusion of voices, as well as their ability to register at both sonic and sense-making levels – which Roland Barthes (1977: 181) called the geno-song and the pheno-song, and which this chapter examines in terms of singing and songwriting voices – offer aesthetically inventive avenues for fan's investigations and yet they also thwart attempts at vocal objectification and, therefore, full knowledge of the documentary subject.

The singer-songwriter film, by definition, grapples with the heterogeneity and multiplicity of the performing voice. The first and still-quintessential example of this subgenre of the music documentary is D.A. Pennebaker's *Don't Look Back* (1967), a film that seems to be more interested in Bob Dylan's hip persona than in his voice. Yet it famously opens with 'Subterranean Homesick Blues' playing extra-diegetically while Dylan holds up cue cards bearing selected handwritten lyrics. This strategy, reportedly proposed by Dylan himself, employs handwritten graphics that link the human body (or at least the hand) to the voice while suggesting the uniqueness of the latter. The cards render the sung lyrics visible as disconnected words, lending emphasis to Dylan's singing voice as a sound because the emphasis on random words confounds lyrical or linguistic sense-making. At moments, Dylan struggles to keep the cards moving as fast as his own voice demands, indicating that Dylan's disembodied singing voice, not his visible body, governs the pace of this film within a film.

Techniques like Dylan's card trick envision the voice as a paradoxical feature of human identity: related to human embodiment while also escaping the body. The sequence manifests the human singing voice's multiple functions as a linguistic medium for conveying poetic meaning, a musical instrument establishing rhythm and pace, and a sonic capacity that exceeds both rhetorical meaning and the discipline of a musical time signature. Strategies that call attention to performing voices may be music documentary's response to the popular culture fascination with singing, the latter evidenced in recent years by the international success of such television programmes as *The Voice, Popstars* and *American Idol.*

They may also represent an 'old media' bid to compete with the immersive potentials that new media formats promise. As Dyson observes, 'the desire for total immersion, as well as the use of audio to simulate this condition, is certainly evident in old media such as cinema' (2009: 3). The goal of these singer-songwriter films is not to capture the sublime spectacular moments within live performances but, rather, to probe the interiority of the performer. In an earlier era of music documentary, visual techniques such as the ideal camera position, well-timed close-ups or innovative editing or cinematography (remembering *Woodstock*'s split screen, for example) were employed to define affective or aesthetically charged moments during a performance. In the contemporary singer-songwriter documentary, by contrast, sound design positions viewers as listeners – eavesdroppers, even – who explore the lives of performers, not merely individual performances. This shift in emphasis to the narrative of the performer's creative and personal life reflects the intensification of a celebrity culture that encourages fans to not just obsess over cherished stars, but also to adopt their personas. Fans may not be able to easily refashion their look to mimic a favourite performer, but they can always sing along.

The three singer-songwriter films this chapter examines, *Amy* (Asif Kapadia 2015), *You're Gonna Miss Me: The Roky Erickson Story* (Keven McAlester 2005)[1] and *Montage of Heck* (Brett Morgen 2015), are compilation documentaries. That is, they are formal hybrids that qualify as Avant Docs (MacDonald 2014). They assemble motley audiovisual sources, blending minimal amounts of performance footage with 'behind the scenes' material in order to define a performer's unique gift and by extension, comprehend their identity.

Montage and *Amy* present their subjects' voices through Joe Gillis-like feats of narration from beyond the grave. *Montage*'s key motif is the reanimation of Kurt Cobain, developed across scenes that animate Cobain's diary drawings and represent him in animated form, culminating with performance footage from the Reading Festival in 1992 where he appeared as a near-corpse in a wheelchair – a stunt intended to mock tabloid reports of his drug addiction. *Amy* edits Winehouse's image and voices – spoken, singing and songwriting – and those of twenty-four friends and associates, orchestrating a conversation that Eisenstein might have called dialectical sonic montage. By contrast, *Miss Me* conveys the struggle of still-living legend Roky Erickson as he attempts to gain control of his voice(s) in the context of a schizophrenia diagnosis and a fractured family.

The *Sunset Boulevard* (Billy Wilder 1950) reference is far from accidental: Wilder's film highlights the license granted to filmmakers who challenge the idea that a single voice must be 'married' to a body to establish cinematic subjectivity. Mary Anne Doane argues that, in narrative cinema, 'the voice must be anchored by a given body' (1980: 76); but recent work by Sarah Kessler (2016) and Patrik Sjöberg (in this volume) shows that documentary audiences accept and may enjoy voice–body mismatches that flout this injunction. Although Gillis's posthumous narration in *Sunset Boulevard* is the most obvious voice–body mismatch in that film, Wilder in fact builds a far more compelling sonic motif around his protagonist's multiple vocal performances. Variations in William Holden's intonation, pace and pitch deliver a range of vocal styles, from the hipster banter he uses to describe Norma Desmond, to the mock-actorly tone he employs when wooing Betty Schaefer, to the nasal barking he deploys when rejecting both women. These manifestations of Gillis's forked tongue find a canny parallel in his authorial production in the form of the two screenplays he is writing with Norma and Betty. Gillis, narrator, character and screenwriter, refuses to be defined by a single voice in life or in death.

Similarly, *Amy*, *Miss Me* and *Montage* explore the multiple voices that emerge from the deep dive these directors make into the archives that inform the aesthetic and cultural meaning of Winehouse's, Erickson's and Cobain's music, performances and personas. All three performers were celebrated as voices of their generation, cast as unique individuals who nevertheless spoke (and sang) for the many. In their musical performances, aspirations towards aesthetic singularity inevitably meet the claims of social collectivity. The lyrics penned and performed by these three musicians speak directly to this paradox, as they present intensely personal feelings of grief, anxiety, depression and anger that resonate with audiences as widely shared affective states.

Amy, *Miss Me* and *Montage* borrow from music, film, television and digital media, proffering formal strategies from covers and remix to selectively stripped music tracks, meme-style interventions recalling *Don't Look Back* and cacophonous blends of sound, all aimed at fully capturing the voice(s) that emanate from the singer-songwriter while also rendering their subjective experiences and star narratives legible.

Vocal dialectics in *Amy*

Released in 2015, four years after Winehouse's death, Asif Kapadia's *Amy* evidenced the singer-songwriter's wide and continuing appeal, earning $4.9 million to become the highest grossing British documentary in history. Winehouse's rise from a precocious teenager to her untimely death at twenty-seven unfolds through a vocal tapestry of narrators whose testimonies are paired with low-resolution home video images. Interviews with Winehouse and twenty-four friends and associates are only heard in voice-over, creating a disembodied dialectical cacophony – as many different speaking voices propose and refute assertions about Winehouse – and producing 'an intimate mode of storytelling' (Andrews 2017: 351). Jennifer O'Meara writes that Kapadia 'avoided the standard "talking head" format in order to keep the visual focus entirely on [Winehouse]. The only time we see the interviewees on screen is if they happen to appear in the scenes with Winehouse' (O'Meara 2016: 2).

Rather than permitting the tabloid press that exploited Winehouse's visual image during her life to determine film viewers' interpretation of her music and personal life, Kapadia works against his own image track in a dialectical manner, engineering a dialogue among multiple voices, including Winehouse's own songwriting, singing and speaking voices and the chorus of those around her. For example, a sequence that features several friends discussing Winehouse's drinking and drug habits is 'answered' by Winehouse's speaking voice saying she is fortunate to have music as an outlet for her depression, belying –or denying – the tale of doom and dependency. This highly sensitive disclosure about depression contributes to the intimacy Andrews finds in the film, as Winehouse's voice, as edited by Kapadia, answers, defends and clarifies speculation, sounding as though she is engaged in conversation.

Ultimately, however, the film presents Winehouse's songwriting and singing voices – not her alternately 'gobby' and introspective voice-over – as the most powerfully authentic expressions of her subjectivity. Winehouse's pianist Sam Beste defines her through her relationship to musical expression: 'She had one of the most pure relationships to music, such an emotional relationship to music. She needed music as if it was a person and that she would die for it.'

Kapadia underscores the primacy of this connection to music by emphasizing the link between Winehouse's singing and songwriting voices, deploying a revised version of the visualization strategy from Pennebaker's Dylan film. Kapadia first screens pages from Winehouse's notebooks containing poetry, lyrics and drawings. Second, he overlays her song lyrics as 'videographic additions' (Andrews 2017: 351) over images of Winehouse in the recording studio and in live performances. These meme-like graphic reminders of her authorial role – the lyrics set out in phrases rather than as random words like Dylan's – function like subtitles, clarifying her mumbles, but also reiterating comments that Winehouse writes about events she has lived through. This graphic embodiment of the authorial, songwriting voice appears in a slightly curling font that mimics Winehouse's handwriting, as made visible in the journals, reinforcing the connections between singing, writing and the body through what seem to be 'handmade' words.

In later scenes, Winehouse's singing and songwriting voices – not her speaking voice – are recruited as a form of narration that plays extra-diegetically. A performance of 'Love Is a Losing Game' plays over scenes of her notoriously tumultuous relationship with Blake Fielder-Civil. In giving the artist the power to tell her story by placing her singing and songwriting voices in dialogue with the images, the director overdetermines interpretations of these events.

Thwarting Kapadia's apparent faith in lyrics as narration, Winehouse's singing voice repeatedly threatens to move beyond sense-making. Her garbled vocal stylings reference affinities with jazz and 'blue eyed soul', repeated in comments made by Tony Bennett and yasiin bey. Winehouse's extra-linguistic vocalizations – her pheno-song – have been heard as racial ventriloquism; as using the voice of another to tell one's own story. Daphne Brooks considers 'the black female voice on *Back to Black* [...] as a non-voice utilized as an object of and a tool to make audible white female desire' (Brooks 2010: 53).

The film dramatizes a moment at which Winehouse withholds her voice(s) from her audience; here, inaudibility asserts desire and demonstrates her demons. On stage in Belgrade in June 2011, a drunk and dishevelled Winehouse literally stumbles her way through the performance, alternatively serenaded and jeered by those in attendance. Bronwyn Polaschek argues that *Amy* 'relies on the archetypal construct of the creative woman as a victim' (2017), however, the editing in this sequence suggests both Winehouse's victimization – she had wanted to cancel the tour but was ignored – and agency – after her virtual non-performance, the tour was cancelled.[2]

In a scene that challenges reports of Winehouse's lack of professionalism due to drug and alcohol abuse, a colleague's voice testifies to her commitment. Producer Marc Ronson's comments play over footage from a studio recording session during the making of *Back to Black*. According to Ronson, Winehouse was disciplined and eager to work. The conclusion of this sequence depicts Winehouse in the sound booth recording vocals. Kapadia alternately adds and removes backing tracks, ultimately singling out Winehouse's singing voice as the focal point of the performance. In a film that treats its subject with some fairly abstract formal practices – using images against the soundtrack and generating a dialectical conversation among twenty-five voices – the film concludes with the proposition that Winehouse's autobiographical lyrics and vocal expression provide the best access to her music-driven subjectivity, reiterating a conventional understanding of the voice as 'the very coincidence of the quintessential corporeality and the soul' (Dolar 2006: 71).

Missing Roky Erickson

Keven McAlester's *You're Gonna Miss Me,* which premiered at the 2005 South by Southwest Festival, foregrounds Roky Erickson's life as a performer, relying on archival footage, interviews, home movies and the sound mix not only to convey the power of Erickson's singing and songwriting voices, but also to suggest the internal voices experienced by the artist, who was diagnosed with paranoid schizophrenia in his early 20s. Schizophrenia remains a condition that medical specialists and those with the condition associate with hearing voices, clinically referred to as auditory hallucinations. Jailed on drugs charges numerous times, he was incarcerated at Rusk State Hospital for the criminally insane in the early 1970s and subjected to Electroconvulsive Therapy (ECT).

McAlester introduces Roky Erickson *as* a voice: an otherworldly medium delivering lyrics that, importantly, move beyond language entirely. The film audience first encounters Erickson in black-and-white kinescope footage of a 1967 Elevators' performance on *Dick Clark's American Bandstand*, where a young Roky appears as a charismatic, screaming shaman. Offering a nostalgic look at a lip-syncing Roky at the peak of his creativity, the film's title sequence plays over the images so that the lyrics of his biggest hit song –'you're gonna miss me, baby' – engulf the screen in a huge white script, adapting the lyrical visualization strategy of *Don't Look Back*, also seen in *Amy.* The clip is followed by an interview with ZZ Top's Billy Gibbons, a friend of Erickson's, who describes Erickson's voice as 'the mystery factor that no one could touch'; echoing a common refrain about Erickson. Mark Deming writes, for example: 'If Roky Erickson had vanished from the face of the earth after the 13th Floor

Elevators released their epochal debut single, in all likelihood, he'd still be regarded as a legend among garage rock fanatics for his primal vocal wailing and feral harmonica work' (2005).

In the two televised performances that open *Miss Me*, Roky's obvious lip-syncing contrasts markedly with the excessive sonic quality of his screaming; it's as if his performing body can't fully re-enact the somatic conditions that produced the scream he must now mimic rather than emit. The tension between his performance and his recorded voice foreshadows Erickson's struggle with the disembodied voices that haunt his internal soundscape. Four minutes into the film, after Gibbons says 'there was no way [the Elevators] couldn't make it', McAlester cuts to the first scene of present-day Erickson, a recluse in his 50s living under the guardianship of his mother Evelyn. The soundtrack produces a rock music version of the sad trombone – a loud falling chord in reverb – as a dishevelled Erickson begins to read a junk mail prize offer aloud. He asks his mother for a Mr Potato Head doll in a childlike whine, which contrasts dramatically with his forceful singing voice, heard as the film cuts to 'Slip Inside This House', an Elevators' song.

In the second present-day scene depicting Erickson, the audience is provided access to his complex subjective experience through the sound mix – his own and the film's. As he enters his Austin apartment, Erickson walks through the space, piercing the silence – and, it's implied, drowning out aural hallucinations – by eliciting sounds from, in turn, a television; an organ that plays 'Silent Night' on repeat; a radio broadcast; a sophisticated mixer that produces ambient noise with eerie vocalizations and recorded harmonica music. After orchestrating this racket, Erickson settles into a chair, telling his mother he is about to take a nap; though visible on screen in the kitchen, she seems unable to hear him over the din. She tells McAlester, 'it's when I turn it off that he wakes up'.

Dissonant soundscapes and mismatched voices and bodies expose the film's concern with the multiplicity of voices Erickson both emits and experiences. Editing juxtaposes Roky as a vocal presence then and now, exploring the variability of his performing voice and suggesting the ways the voice is not controlled by its ostensible owner. The legal dispute at the centre of the narrative – Erickson's youngest brother Sumner seeks to replace Evelyn as his guardian – exposes the equivalencies the entire family makes between Roky's performing voice, his self possession and his subjectivity. Evelyn states that playing guitar is 'the only time [Roky] seems completely happy'. Sumner couches his expectations of Roky's success in recovery in terms of 'getting him up singing again'. After the courts reassign guardianship to Sumner, Roky begins medication and does indeed begin singing again – on camera, to the filmmaker with an acoustic guitar – yet he remains incapable of contributing to his own narrative in his speaking voice.

The uncanny vocal multiplicities that surround Erickson cannot be harnessed to make him whole enough to coherently author his life story; interviews with others must take on that duty. A sequence on his post-incarceration career in the 1980s reveals that his band Bleib Alien merged anti-communist paranoia (Sputnik) with references to space aliens in an anachronistic voicing of his paranoid delusions. The film's presentation of voices that are intimate and screaming, vocalized and hallucinated, may cause the listener to question even those musical performances whose claims to authenticity rest on autobiographical information – as is strongly implied in both *Amy* and *Miss Me* – because voices can in fact reflect a profound disassociation from subjectivity altogether. One scene draws a connection between Erickson's lyrics and poetry – 'voiced language', according to Don Ihde (2007: 196) – through footage from Austin State Hospital dated 1986 in which Erickson recites a bone-chilling spoken word piece about domestic abuse, 'I Know the Hole in Baby's Head'. The implication of still-unacknowledged child abuse suggests the possibility that Erickson's powerful screaming voice in young adulthood might have been his response to *not* having his voice heard when it mattered.

The poignant title song and Elevators' signature hit, *You're Gonna Miss Me*, thus becomes a complex utterance. It appears at first listen to be a defiant tale of romance on the brink of doom. But by the end of the film, the song functions as a potentially ironic statement about a performer's presence, or lack thereof. One line reads, 'You're gonna look around *in your mind*, girl/you're gonna find that I'm gone' (emphasis added). By referencing the listener's mental fantasy of the singer, the line in the context of the documentary may allude to the existentially ephemeral status of the performer (not just the lover), to the lingering desire of the spectator after a performance, and/or to the mourning of fans when performers die. Whereas Kapadia uses multiple interpenetrating voices in *Amy* to suggest that music is the key to her subjectivity, in *You're Gonna Miss Me*, McAlester's sound design suggests that not all voices are real and that even the most emotive performance might represent the subject's displacement. Even when he is standing right before your eyes, his ferocious wailing penetrating your listening body, you might miss Roky entirely.

Montage of denial

The title of Brett Morgen's Kurt Cobain documentary, *Montage of Heck* (2015), derives not from a major album or hit song, but instead from a low-fi mix tape where Cobain recorded his singing and speaking voice, ambient sounds and sonic experiments. The credits grant Cobain authorial status, listing him as a sound designer, along with career professional Cameron Frankley. The tape exerts authority over the film's discourse as a source of Cobain's intimate musings and not-for-public consumption experiments which, along with home movies and interviews recorded for the film, promises access to the performer's subjectivity.

Cobain himself, however, rejects the possibility that the tape – or any performance – can provide a singular voice through which to understand the artist. In his speaking voice, the singer-songwriter repeatedly deflects questions about his music, telling interviewers that he is more interested in hearing fans' interpretations. This gesture explicitly acknowledges that performing voices, in both rhetorical and sensorial registers, are shared between performer and listener. As with *Amy* and *Miss Me*, voices emerge as multiple, intersubjective, potentially other and resistant to ownership.

Harnessing Cobain's opposition to the manicured production values of the music industry, *Montage* refuses to give the audience high-fidelity, full-length Nirvana performances. In fact, instead of providing fans with large quantities of Cobain and Nirvana performance footage, the film frequently recruits the voices of other people in interviews and, more unusually, in covers of Cobain's music. Indeed, the film opens with images of Cobain in diapers 'set to a lullaby-sounding cover of Nirvana's "All Apologies" performed by Rockabye Baby!' (Garofalo 2015). Perhaps the most dramatic example of the film's engagement with other voices occurs with the dirge-like rendition of 'Smells Like Teen Spirit' by the Scala & Kolacny Brothers' Belgian women's choir. Here the soundtrack moves far afield from the grunge aesthetic in order to trace the extensive influence of Cobain's songwriting voice.

The primary conceit of *Montage* is the reanimation of Cobain's voice, in the film's discourse and through the cover songs, as well as his body. Editing image and sound in counterpoint links his songwriting and singing voices with a high degree of reflexivity. Jennifer O'Meara describes an example of this device:

> The closest *Montage of Heck* gets to showing his corpse is through the repeated use of two images of Cobain *pretending* to be dead. [...] The second time one of these faux-corpse images is shown, it's accompanied by Cobain saying in voice-over that people seem to want him to die because that's the classic rock star story. (2016: 2)

Here, Cobain appears dead but his voice lives on, which performatively confirms the classic rock star story.

The film teases out the way Cobain's singing and songwriting voices express his physical state. 'Cobain prized scatological imagery, eviscerating vocals, and unintelligible lyrics as a means to signal the "realness" of his art – that is, his art came from the "gut"' (Wood 2011: 332). Cobain's obsession with ingestion, expulsion and oral projection – linked to a painful stomach disorder suspected as contributing to his heroin addiction – is documented in writings and drawings in his diary, images of which are frequently displayed on screen and animated.

Vocal performance assumes a gendered aspect in the film, underscoring the connection between Cobain's voice and body. The film highlights his interests in shattering gendered assumptions, evident, for example, in his pregnant man drawings. The flexible play of genders through vocal registers, as well as the fraught aspects of the performer–fan relationship, are apparent in home movie footage depicting Cobain and his wife Courtney Love. They perform an exaggerated and ironic reading of a fan letter critical of Courtney: Love adopts the aggressive voice of 'Sassy' while Cobain performs as a ventriloquist's dummy whose voice is not its own.

Montage culminates with the assertion of Cobain's screaming voice, for which he, like Erickson, was famous. Using the same technique that Kapadia employs in *Amy*, subtracting audio tracks and leaving Cobain's voice unaccompanied, *Montage* concludes its credits sequence with a prolonged track of 'Smells Like Teen Spirit' in which Cobain repeatedly screams 'a denial'. This gesture may seem to reassert the singularity of the singer-songwriter's voice, distinguishing it from those who covered his music, yet it also foregrounds the way Cobain's repetitive screaming moves between sense and sensory expression.

Screaming like this not only takes singing beyond language, it also invites the collapse of norms around social interaction and may threaten subjectivity itself. Nicollete Rohr considers the 'screamscape' of Beatlemania a site of transgressive, gendered rebellion (2017: 4). Patrick Burke reports that female screamers are linked to sexual impropriety, whereas male screamers are 'depicted as a potentially violent mob' (2013: 17). Burke pursues the existential implications of screaming as an activity that is 'both personal and collective [...] Screamers engage at a visceral level with the question of what it means to maintain one's identity within mass society' (20).

When fans interact vocally with performers' voices (whether live or recorded) by singing or screaming along, they use this multiplied vocal capacity to engage intersubjectively. The scream is a voiced intervention that, for Michel Chion, represents 'a rip in the fabric of time' (1999: 77); for Rhiannon Harries, a scream 'harbour[s] an affective temporality outside the linear progress of rational language or history' (this volume: 280). Neuroscientists suggest that the biochemical effects of group vocalization such as singing and screaming enable people to enter social flow states (Keeler et al. 2015). By foregrounding the scream, McAlester and Morgen's films invite fan recall of, or participation in, a social flow that transcends linear time and may dispense with the notion of individual subjectivity completely.

Conclusion

The suggestion that the human voice functions as a central concern of music documentaries seems embarrassingly obvious. Indeed, Justin Horton has made a more radical assertion that 'despite the still-prevalent bias toward the visual among spectators, critics, and scholars, [sound films] hinge more often than not on the voice' (2013: 4). And yet the dearth of scholarship on the music documentary – as Michael Saffle has noted (2013: 43) – much less on the role the human voice plays within it, demands further scholarly examination.

This chapter concludes with three observations. The first is that the singer-songwriter portrait – and, I would argue, all music documentaries – speak to the desire of fans, filmmakers and perhaps even the musicians themselves, to understand the performing voice. That voice is intriguing, enigmatic and privileged precisely because it conveys lyrical sense and promises to reveal subjective interiority and yet also moves beyond linguistic meaning and offers avenues for the deconstruction of subjectivity. Recent singer-songwriter portrait documentaries thus reveal a paradox: in spite of or, perhaps, because of, their archive-enhanced aesthetic of multiple voicings, these films reveal that it is impossible for the 'owners' or admirers of such voices to completely define or fully possess them. In other words, these films challenge the conventional understanding of the voice as 'the very coincidence of the quintessential corporeality and the soul' (Dolar 2006: 71).

Second, examining music documentaries – and, as this collection as a whole argues, all documentaries – demands a departure from existing practices of documentary scholarship. Bill Nichols's assertion that, when considering documentary, 'questions of voice are [...] not meant entirely literally' (2017: 48) must be revisited in order to address the role of voices as they produce speech, music and sound as an integral aspect of documentary representation.

And finally, recent singer-songwriter documentaries emphasize the human voice and offer the potential for a specific type of engagement in social flow. Norie Neumark argues that fiction films that privilege the voice create 'a sense of interiority, involvement, and intimacy with the audience' that enables the voice to 'tie the audience into the narrative' (2010: xxiii). Harnessing narrative but rejecting the virtual spaces or 3D techniques that are touted as immersive cinema, singer-songwriter documentaries invite audiences into the sonic multiplicity of and, potentially, the collective flow associated with human voices.

These enveloping sonic encounters with performing, authorial, writing, informal and alternative voices mediate and enhance the backstage tale, attempting to explore the performer's interiority rather than recreate the spectacle of the live performance. Dyson writes that 'sound – by virtue of its curious ontology – performs, conceals and negotiates different technocultural narratives at different historical junctures' (2009: 10), and the revitalization of the singer-songwriter music documentary, with its obsession with voice, at this historical juncture represents a paradoxical desire for engagement with and erasure of subjectivity. Reconsidering the music documentary as a site of vocally and sonically immersive spectatorship argues for its continuing relevance to the study of film form and offers one avenue for considering broader questions regarding the human desire to understand the subjectivity of others.

Notes

1 Erickson (b. 1947) may be the least well known of the three. His Texas band, the 13th Floor Elevators, is credited with inventing psychedelic rock before The Doors popularized it, and lead singer Erickson is recognized for his screaming wail, which influenced his Texas-born contemporary Janis Joplin.

2 The film gives the mistaken impression that Winehouse did not sing at all, which overstates her withholding the voice as an expression of her displeasure.

References

Andrews, H. (2017), 'From Unwilling Celebrity to Authored Icon: Reading *Amy* (Kapadia 2015)', *Celebrity Studies*, 8 (2): 251–254.

Barthes, R. (1977), 'The Grain of the Voice', in *Image, Music, Text*, trans. S. Heath, 179–189, New York: Hill & Wang.

Beck, J. with V. T. Ament (2015), 'The New Hollywood, 1981–1999', in K. Kalinak (ed.), *Sound: Dialogue, Music and Effects*, 107–132, New Brunswick, NJ: Rutgers University Press.

Brooks, D. A. (2010), '"This Voice Which Is Not One": Amy Winehouse Sings the Ballad of Sonic Blue(s)Face Culture', *Women & Performance: A Journal of Feminist Theory*, 20 (1): 37–60.

Burke, P. (2013), 'The Screamers', *Daedalus*, 142 (4):11–23.

Cagle, C. (2012), 'Postclassical Nonfiction: Narration in the Contemporary Documentary', *Cinema Journal*, 52 (1): 45–65.

Chion, M. (1999), *The Voice in Cinema*, trans. C. Gorbman, New York: Columbia University Press.

Deming, M. (2005), 'Overview' in *I Have Always Been Here Before: The Roky Erickson Anthology*, AllMusic.com, retrieved 2018-1-8.

Doane, M. A. (1980), 'The Voice in the Cinema: The Articulation of Body and Space', *Yale French Studies*, 60: 33–50.

Dolar, M. (2006), *A Voice and Nothing More*, Cambridge, MA: MIT Press.

Dollar, S. (2018), 'Sundance 2018 Preview', Billboard.com, retrieved 2018-1-8.

Donnelly, K. (2015), *Magical Musical Tour*, New York: Bloomsbury.

Dyson, F. (2009), *Sounding New Media: Immersion and Embodiment in the Arts and Culture*, Berkeley, CA: University of California Press.

Gallo, P. (2015), 'Banner Year for Music Docs,' *Variety*, 329 (17): 121–122, 124.

Garofalo, A. (2015), '"Montage Of Heck": 6 Music Moments, Unreleased Tracks and Live Performances Featured in the HBO Kurt Cobain Documentary', *International Business Times*, 29 September 2015. Available online: http://www.ibtimes.com/kurt-cobain-home-recordings-be-released-montage-heck-soundtrack-director-teases-2119250 (accessed 15 April 2018).

Horton, J. (2013), 'The Unheard Voice in the Sound Film', *Cinema Journal*, 52 (4): 3–24.

Ihde, D. (2007), *Listening and Voice: Phenomenologies of Sound*, 2nd edn, Albany, NY: SUNY Press.

Kapadia, A. (2015), *Amy*, Film.

Keeler, J., E. Roth, B. Neuser, J. Spitsbergen John, D. Waters and J. M. Vianney (2015), 'The Neurochemistry and Social Flow of Singing: Bonding and Oxytocin', *Frontiers in Human Neuroscience*, 9: 518.

Kessler, S. (2016), 'Puppet Love: Documenting Ventriloquism in Nina Conti's Her Master's Voice', *Camera Obscura: A Journal of Feminism, Culture, and Media Studies*, 31 (92): 61–91.

MacDonald, S. (2014), *Avant-Doc: Intersections of Documentary and Avant-Garde Cinema*, London: Oxford University Press.

Morgen, B. (2015), *Kurt Cobain: Montage of Heck*, Film.

Neumark, N. (2010), 'Introduction', in N. Neumark and R. Gibson (eds), *Voice: Vocal Aesthetics in Digital Arts and Media*, xv–xxxii, Boston, MA: MIT Press.

Nichols, W. (2017), *Introduction to Documentary*, 3rd edn, Bloomington, IN: Indiana University Press.

O'Meara, J. (2016), 'The Troubled Production of Troubling Images in the Contemporary Music Documentary', Unpublished paper presented at Music and Visual Cultures Conference, Maynooth University, 21–23 July.

Polaschek, B. (2017), 'The Dissonant Personas of a Female Celebrity: Amy and the Public Self of Amy Winehouse', *Celebrity Studies* (May 2017): 1–17.

Rohr, N. (2017), 'Yeah Yeah Yeah: The Sixties Screamscape of Beatlemania', *Journal of Popular Music Studies* 29(2). doi:10.1111/jpms.12213

Saffle, M. (2013), 'Retrospective Compilations: Redefining the Music Documentary', in R. Edgar, K. Fairclough-Isaacs and B. Halligan (eds), *The Music Documentary: Acid Rock to Electropop*, 42–54, New York: Routledge.

Wood, J. (2011), 'Pained Expression: Metaphors of Sickness and Signs of "Authenticity" in Kurt Cobain's "Journals"', *Popular Music*, 30 (3): 331–349.

5

'The Voice' and Sound in Indian Documentary Film: Listening to Surabhi Sharma's *Bidesia in Bambai* (2013)

Deborah Matzner

Scholarly discussion of documentary film and 'the voice' tends to focus on the politics of representation – who gets to speak for whom and how (see, for example, Nichols 1994; Ruby 1991). Given the power dynamics that often organize the relationships between the makers of non-fiction film and their subjects, these are crucial questions to ask in any documentary tradition. They are particularly apt in postcolonial contexts like those that structure Indian documentary film practices, where the question 'can the subaltern speak' must be asked of all elite representation that seeks to portray vernacular perspectives (Spivak 1988). Indeed, in India, where documentary film typically is used by an activist intelligentsia for social critique focused on the experience of the non-elite, these are questions constantly asked of non-fiction film, by no one more than the filmmakers themselves.

Nevertheless, 'the voice' suggests not only a representational phenomenon, but also a sonic one. Non-fiction films often re-present the vocalizations of their subjects: shouts, testimony, song, life history, the wordless sounds of human presence. Sound is crucial to the sensory experience of everyday life, as well as to the mediation of social and political sentiment, all of which non-fiction film, especially Indian documentary, tends to document. This chapter contributes to the recent 'sonic turn' in the study of social life (see Erlmann 2004; Faudree 2012; Feld and Brenneis 2004; Feld et al. 2005; Hirschkind 2006; Sterne 2003) and of documentary film (MacDougall 2006) by expanding the idea of 'the voice' beyond considerations of text, representational authority and point of view. I hope to illustrate that by attending to voicing as a physical and social phenomenon, we can gain purchase on some cultural specificities that inform relationships among voice, subject, medium and community – both in Indian documentary and beyond.

The Indian documentary tradition presents a rich context in which to situate such an analysis. Recognizing the power of 'the voice' for political address and community formation in the Indian context, many contemporary documentary filmmakers are experimenting with ways to re-represent vocal practices in order to harness, comment upon and at times intervene in the affective power of sound, negotiating their own relationships to various voices and their intents. In what follows, I explore such innovations in the work of Surabhi Sharma, an important Mumbai-based documentary filmmaker whose 2013 feature-length film displays an ethnographic interest in the politics of song for the constitution of identities and the formation of communities. Her film *Bidesia in Bambai* (2013) explores connections between sung performance in the Bhojpuri language and affective community among the Bhojpuri-speaking migrants from northern India who occupy peripheral spaces in Mumbai. *Bidesia* exemplifies the importance of song and sound for the mediation of affect in Indian documentary and everyday life, as well as hinting at their particularities in this specific context. Moreover, the sound design choices that Sharma has made not only to represent but also to trouble 'the voice' of Bhojpuri migrants exemplify the complex, self-aware politics of representation in contemporary Indian documentary film.

Sound and the voice in India – anthropological approaches

In order to begin to understand the operation of sound and 'the voice' in Indian everyday life and the non-fiction filmmaking practice that seeks to represent it, it is helpful to draw on insights gleaned from the anthropology of sound and of 'the voice'. The anthropology of sound takes as its subject the sociality and materiality of sound. Analysing sound's *materiality* involves attending to its physical qualities and capacities. Consider, for example, the ways in which sound relates to touch. It physically interacts with the perceptive membranes of the body – the eardrums, as well as outer ear, and, at sufficient volume, the skin and other organs (see Connor 2001). Sound has the capability to transcend boundaries, marking and exceeding physical barriers and locations and indicating presence and activity, albeit in fleeting, evanescent ways (Labelle 2010; Samuels et al. 2010). All of these capabilities of sound are useful for its social effects, which can involve the shaping of identities and communities (Keil and Feld 1994: 18; Revill 2000). But attending to the *sociality* of sound cautions us against universalizing impulses in the theorization of how sound works. Recognizing the deep cultural and religious specificities of theories and practices of mediation, communication and embodiment that fundamentally shape people's experience and practice of audition and sound-making, anthropologists attend ethnographically to the different ways sound's material characteristics and social capabilities are taken up in various contexts (Erlmann 2004: 3; Hirschkind 2006; Stoller 1989; Taussig 1993).

These points encourage us to recognize the affective and political power of 'the voice' on screen and off, as well as the cultural specificity of not only its sonic operation, but also its metaphorical and social aspects. From anthropological perspectives, the voice is a liminal boundary, expressive and impressionable, between the body and the social.[1] Common-sense notions of people's physical voices frame them as indexical of essential qualities of the individual, when those qualities are more correctly attributable to *habitus* than to nature. As anthropologist Amanda Weidman put it, 'voice is a crucial site where the realms of the cultural and sociopolitical link to the level of the individual, a site where shared discourses and values, affect, and aesthetics are made manifest and contested through embodied practice' (2014: 38).

Metaphors of the voice as representation rely on particular ideologies of mediation, community and personhood, which can differ by culture. The cultural specificity of mediation has been suggested by scholars of South Asian visual culture analysing *darshan* and *nazar*, which are extrusive, haptic, corporeal and interactive modes of seeing and being seen in the contexts of the visual interaction in romantic love and in Hindu religious devotion and political spectatorship (see Bhatti and Pinney 2011; Pinney 1997; Taylor 2002; see also Matzner 2014). These forms of vision 'provincialize' dominant Western conceptions of vision as a disembodied sense capacity of the rational modern individual. Work on sound and audition in Hindu and Muslim religious contexts has pointed to particular philosophies of the divine, mediation and sound that trouble universalizing descriptions of sound's operation (Beck 1993, c.f. Gade 2004). Local philosophies of personhood often defy the presumption of 'the individual' as a singular, essential entity available for representation (see Daniel 1984; Marriott 1990; Pinney 1997; Trawick 1990). Moreover, in terms of political representation, in India, the community as voting block often operates as the more salient atom of the democratic process than the individual (see Appadurai

1993; Kothari 1989). All of these insights suggest challenges to notions of 'voicing' in its sonic and metaphorical aspects as a universal process.

Sonic and representational practices in Indian documentary film

In India, the documentary film has its origins as a colonial tool for social persuasion, although today its sonic practices and modes of address have been adapted to the goals of social analysis and critique. Documentary was key to the postcolonial project of Nehruvian national development. The government's Films Division (FD), established after Independence as a repurposing of the colonial bureau Information Films India, had a near monopoly on documentary production for decades. Commercial cinemas were required to screen FD shorts prior to feature films. These documentaries addressed citizens with paternalistic 'voice of God' narration accompanied by folk music, celebrating village practices and the modernizing projects of the nation (see Roy 2002). In style and mode of address the FD documentary and Indian non-fiction media more generally owed a great deal to pioneer of British documentary John Grierson. James Beveridge, Grierson's close associate, worked as a consultant to many post-Independence Indian media bureaucracies, including FD. Griersonian documentary involved a hierarchical model of communication, in which progressive but never truly revolutionary ideological content is 'passed down from a bureaucratic elite, via the documentary film, to the public' (Aitken 2006: 523). The authoritative voice of the FD documentary was, therefore, that of the postcolonial Indian state.

This changed briefly in the late 1960s and early 1970s under the directorship of Jean Bhownagary, who oversaw a cohort of filmmakers influenced by European and Russian experimental film. For example, one FD director at this time was Pramod Pati, whose films employed visual and sonic experimentation. Pati physically drew directly onto his film stock to create audio and visual effects and used pixilation and rapid visual and sonic montage. He coupled this formal experimentation with anti-authoritarian content. To give a striking example, Pati's film *Explorer* (1968), a rapidly paced visual and audio montage depicting the curiosity and scepticism of Indian youth, included a brief slate reading 'f*ck censorship!' With such films, the politics of the documentary were antithetical to those of the state, a stance echoed in experimental sonic practice.

The suppression of civil liberties during Prime Minister Indira Gandhi's Emergency Rule (1975–7) and the dwindling of FD's funding helped to prompt the emergence of the independent documentary film movement in which filmmakers today participate. Having once conveyed the voice of the state, Indian documentary film was transformed in the 1970s into the voice of Left-liberal activism. Indian documentary became primarily associated with social critique and activism on the part of the intelligentsia, who were committed to representing the experience and oppression of the non-elite in order to critique the failures of Indian modernity. As filmmaker and cultural commentator Paromita Vohra (2011) has pointed out, this has meant that Indian documentary tends to eschew self-reflexivity in order to focus on social problems, seeming to convey the 'voice of the people' while remaining an expressive tool of the relatively elite. Sonic practices in this advocacy mode of political documentary tend to involve conveying the political power of song and protest in order to remediate it faithfully and thus rouse documentary's viewers into affective affiliation with the perspectives of the oppressed.

Surabhi Sharma's films, while deeply indebted to the dominant form of socio-realist Indian documentary, depart from these practices. Sharma belongs to a younger generation of Mumbai-based documentary practitioners who have been inspired by the political and activist mode of independent documentary film in India, but are seeking to innovate on it (see Vohra 2011; Wolf 2002). Early in her career, Sharma worked with Anand Patwardhan, a pioneer of independent Indian documentary film whose work exemplifies, and has helped to render normative, the realist political mode of documentary that advocates for the oppressed and seeks to convey their voices. Across four decades, Patwardhan's films have presented among the most powerful depictions of the rise of Hindu nationalism, the exploitation of the urban poor and rural indigenous groups, and the many failures and injustices of the Indian state to uphold its social and secular ideals. Patwardhan has been influential as a leading activist for creative freedom and the right to dissent, as well as a mentor to many younger filmmakers, including Sharma.

In addition to her work with Patwardhan, Sharma is also influenced by her training in anthropology. Using the film camera like a field notebook, she registers the experience of on-the-street, in-camera analysis, favouring a loose, open-ended format that is true to the contingency of everyday life and to her inductive research process rather than the didacticism or closure of other forms of documentary. As the film scholar and curator Nicole Wolf has written,

'Surabhi Sharma's progression and experimentation attempts to work alongside the poetics of the socio-realist form which have become conventionalised and institutionalised, and to arrive at other politics from there' (2002: 127). In particular, in her recent work, Sharma is experimenting with both sound and mode of address. She does this in order to present the perspectives and sonic practices of her non-elite filmic subjects without aligning the politics of her film with those of her subjects.

Sound and 'the voice' in and of *Bidesia in Bambai*

Sharma's 2013 feature-length project, *Bidesia in Bambai*, roughly translatable as 'Migrant in Mumbai', focuses on Bhojpuri music in Mumbai. The Bhojpuri language is associated with a large community, mostly made up of men, who have left their homelands and families in Bihar and Uttar Pradesh in successive and ongoing waves of migration, which began with military enrolment as early as the fifteenth century and continued at an accelerated pace in the 1800s with the abolition of slavery and the need for low-wage labour, through the present (Niranjana 2006). Those who have arrived in Mumbai occupy spaces of marginality in many ways. As newcomers, they take up undesirable and dangerous positions in construction, sweatshop-based manufacturing and as taxi- and rickshaw-wallahs. They tend to dwell in the unruly outskirts of the city, where their homes, officially illegal constructions, are subject to a continuous cycle of destruction and reconstruction. Adding to their precarity, for more than five years local Maharashtrian Hindu chauvinist groups have specifically targeted the Bhojpuri community for discrimination and attack as part of a project to use the streets and the political sphere as arenas in which to assert Maharashtrian claims to the city (see Hansen 1999).

Bidesia explores the Bhojpuri music scene in Mumbai by portraying live performances, recording sessions and interviews with singers. At times, the film's audio track faithfully conveys the sonic details of Bhojpuri music in its live and recorded incarnations. These details include sonic effects introduced by the limits of the human voice in contexts of live vocal performance and the technological qualities of imperfect amplification and reproduction. At times the film also depicts the embodied auditory experience of listening to music performed live in the midst of a cacophonous city. At other times, Sharma's sonic choices serve more analytical than indexical aims. In these cases, Sharma uses non-diegetic sound to distance the viewer of her films from both the subject position of performer and audience member of live Bhojpuri musical performances. She does this in order to open up space for critical reflection on the part of her viewer, critiquing rather than reproducing the affective capabilities of sound and 'the voice'.

Those capabilities are profound. Bhojpuri popular music asserts powerful counter-claims to Bhojpuri belonging in the face of the multiple peripherality of the community. As Sharma put it in an interview, she was inspired to focus on Bhojpuri music in part because she realized

> the migrant pushed to the edges, vulnerable in his economic and social life, was not the story [I found] unraveling in the musical sphere. Instead there seemed to be an assertion of the migrant's claim to the city, histories of love and longing, and a political charge that lay claim to the city. (Sharma 2007)

Not only does the content of Bhojpuri music celebrate the tenacity of migrants despite their marginality, the industry that produces and circulates Bhojpuri songs thrives in marginal and improvised conditions, through informal practices categorized as 'piracy' by mainstream industries and intellectual property laws (see also Manuel 1993: 78; Sundaram 2015). *Bidesia* depicts a number of outdoor concerts held on the rubble fields of recently destroyed buildings and attended by masses of young male migrants. Performers sing of devotion to the Bhojpuri mother goddess and of the sexual frustrations of young men far from their wives and homes. Their tunes aestheticize the migrant status and reframe it as heroic rather than marginal. In depicting the concerts, the director turns her camera not only on the musicians, but also on the audience members, many of whom record performances on their mobile phones, which they then use to play and exchange songs. This practice resonates with a common lyrical leitmotif within the songs themselves that celebrates the mobile phone as a technology that transmits intimacy and connects migrants to home across otherwise prohibitive distances (see also Fernandes 2012).

In the documentary, several performers serve as the key characters whose musical practices grant the viewer insight into the various ways in which music functions as a mediator of affective politics and community sentiment in the spaces of Bhojpuri cultural activity. These performers include a teenager named Kallu, whose specialty is the raunchy *masala* songs filled with thinly veiled double-entendres that substitute mangoes for breasts and roosters for penises. In neighbourhood gatherings, Kallu sings to throngs of men while his female backup dancers whip up the crowd's sexual energies. His songs express a subaltern male sexuality that aggrandizes the Bhojpuri migrant male, countering his ostracizing and vilification by Marathi chauvinists, subverts high class and caste norms of propriety and objectifies the female body. At times, female singers participate in this, echoing the salaciousness of Kallu's performance.[2]

These tendencies are exemplified in one of Kallu's performances. Sharma films from behind the performers onstage, the recorded audio track conveying the blown-out bass notes of overloaded speakers amplifying sound at high volume. Kallu addresses the crowd, the meaning of his husky, youthful speech translated from Bhojpuri to English in Sharma's subtitles: 'All those from Bihar and U.P., raise your hands!' Sharma stays onstage as the crowd roars and Kallu and his female partner return to a salacious song. 'I poke my beak in your nectar cup', Kallu belts naughtily. 'You are eyeing my mango basket', his partner sings in response, shaking her chest vigorously. Then, suddenly, the audio track becomes nearly silent as the image of the performance continues. The audio is replaced by the sound of a cell phone ringing. Then, Sharma's visuals cut to a medium shot of a man in the audience answering his phone, the indexicality of sound and image jumping from the subject position of the stage to that of the audience as if to highlight the theme of mobility and communication across distance that the phone symbolizes in Bhojpuri lyrics.

Figure 5.1 Kallu on stage. Still from *Bidesia in Bambai* (Surabhi Sharma 2013). Courtesy Surabhi Sharma.

In addition to Kallu, another performer who features several times in the film is Vijay Lal Yadav, who performs a style of song called *birha* (separation or longing), associated with the Yadav caste, which occupies a lowly 'Other Backward Caste' position. Many of the *birhas* that Yadav performs in the film articulate a strong, sexualized, masculine Bhojpuri identity as do Kallu's performances. But rather than singing with a woman as Kallu did, Yadav sings *as* a woman, or rather a titillating caricature of one, mimicking the wife left alone in the village. In one scene, Sharma films from onstage at a large outdoor concert as Yadav sings, swaying his hips in a burlesque of a yearning woman and raising his voice into a coy falsetto. Sharma's subtitles translate his words: 'Oh my beloved. If you go and live in Bambai, then who will marry me? Oh, the pain. It is hurting me, my dear. Oh, the pain.' Yadav's lilting, high-pitched voice and swaying hips suggest not pain but sexual pleasure.

He interrupts his singing to address the crowd in an enthusiastic spoken register. 'In English they use the word "pain". In Hindi too they have the word "dard,"' he begins, breathing hard with exertion from his performance. 'But in our Bhojpuri language, there is a specific word to describe every kind of pain. Like for the throbbing head, the sore back, the aching heart. [...] Truth be told, there is no language like Bhojpuri.'

Here, quite explicitly, Yadav aestheticizes the longing and suffering of the Bhojpuri community. The yearning of the wife left behind in the village becomes

Figure 5.2 Yadav imitating the lonely village wife. Still from *Bidesia in Bambai* (Surabhi Sharma 2013). Courtesy Surabhi Sharma.

a source for sexual titillation, its caricature in voice and bodily gesture an object for the performance of heteronormative Bhojpuri masculinity in defiance of the homosociality of the nearly all-male migrant community. The capacity of the Bhojpuri language to express a poetic spectrum of pain becomes a source of pride, proof of superior cultural heritage. The multiply peripheral and multiply mobile Bhojpuri community is invited to unite in its celebration of the suffering male migrant rendered heroic. At this point, Sharma's film does not intervene, faithfully depicting from the perspective of the stage the sounds of speech and song that celebrate *Bidesia* and *Bhojpuria.*

Bidesia in Bambai culminates in a huge religious gathering featuring Yadav as headliner. Here Yadav trades his role as the lonesome wife for that of the Hindu devotee seeking to invoke Bhojpuri Hindu religious ardour. The occasion is the celebration of *Chhath Puja*, an annual holy day worshipping the mother goddess, a major event in the Bhojpuri Hindu sacred calendar. Unlike the other performances depicted in the film, the *Chhath Puja* does not take place in the marginal Bhojpuri neighbourhoods that skirt the edges of the city. Instead, the massive event is held on Juhu Beach, an important recreational attraction situated in a wealthy area known for its celebrity denizens. Juhu Beach is also the site for the annual *Ganesh Chaturthi* festival, which has become a politicized performance of Maharashtrian Hindu nationalist dominance of the city. The Bhojpuri celebration offers a competing claim to belonging.

Figure 5.3 Yadav on the Chhath Puja stage. Still from *Bidesia in Bambai* (Surabhi Sharma 2013). Courtesy Surabhi Sharma.

On a huge stage, Vijay Lal Yadav sings, his raspy voice connoting both exhaustion and religious ardour: 'I'll give you the Mother Goddess's mobile number: chant, penance, meditation, frugality, prayer, worship, devotion. These are the numbers to call the mother goddess!' Sharma's camera moves between a shot of the stage on which Yadav performs in front of a digital image of himself, the rapt faces of the crowd and the tall cranes of camera operators capturing the scene. The image and voice of Yadav, a low-class migrant from an oppressed minority, are projected in gigantic proportions and amplified in ways that must send echoes across this populous and upscale neighbourhood. The crowd itself acts as a manifestation of the defiant presences of the Bhojpuri community.

Mediation is highlighted in this performance as image and sound are recorded and transmitted, and the act of filming and remediating becomes a part of the performance. As Sharma put it at a screening of her film, here the Bhojpuri 'are not just making a production of their presence … they're making a production of their production' (Sharma 2007). What is celebrated here is identity across distance, mediated by technologies like the loudspeaker – a mobile, migratory Bhojpuri belonging, characterized rather than weakened by dislocation.

As the scenes discussed above begin to suggest, Sharma has chosen to remediate Bhojpuri music in a way that differs significantly from how Indian documentary filmmakers in the socio-realist activist mode typically portray the sonic practices of subcultural groups. Indian documentary usually harnesses the sound of protest and presence – the testimonies of individuals, the raised voices of crowd protestors, the songs of the oppressed – in order to remediate and amplify the political claims made by those it depicts (see Matzner 2014). In distinction, Sharma often distorts and interrupts the sound of Bhojpuri vocal performance on her film's audio track. Although in the example of Kallu's performance above, these sound design choices can be seen to shift the perspective to that of an audience member, at times the sound is more experimental, its diegetic referent less clear. Sometimes Sharma fades out the sound on songs and fades in a feedback noise or hum reminiscent of the ringing in one's ears after a loud concert. Sometimes the sound stops entirely as the visuals continue.

Indeed, Sharma has said she has several intentions for her sound design. To a certain extent, her sonic effects are meant to convey the phenomenological experience of listening to performed and mediated sound more directly than a 'clearer' audio track might. As an audience member at a loud event, one cannot hear clean audio, and may receive the poetics of enthusiasm and the volume of mediation rather than the semantic meaning of sung or spoken words.

Sometimes the sounds of interference on the film's audio track, while not strictly diegetic seem to connote the improvised and compromised conditions of amplification of many Bhojpuri concerts, where power outages and technical glitches are common or the muffled sounds of imperfect recording via mobile phone. In this sense, the film's audio track emphasizes what Brian Larkin has called the 'pirate aesthetic' of media practices in imperfect conditions (2004). 'Piracy', Larkin concludes, 'creates an aesthetic, a set of formal qualities that generates a particular sensorial experience of media marked by poor transmission, interference, and noise' (2004: 291). What from the perspective of corporate media industries would appear to be piracy constitutes the normal mode of operation of the Bhojpuri music industry and helps contribute to the unique characteristics of its sound, which Sharma's sound design choices often convey and emphasize.

At other times, however, the dips in diegetic sound on *Bidesia*'s audio track function to distance the film audience from the sensorial experience of being present at the Bhojpuri performances the film depicts. Sharma has acknowledged that she conceived of her sonic distortions as a Brechtian or postmodern anthropological effort to call attention to the fictive nature of the film-as-artefact and to question the authority of the filmmaker (Sharma 2014). In this sense, her sound design practices echo the experimental interventions of filmmakers like Pramod Pati, whose interruptions on the audio track of his films reinforce the disruption of the authoritative voice of the documentary film and of the state for which the filmic genre had been a mouthpiece. This is not the typical documentary voice as voice-of-the-people that tends to characterize the independent documentary film in India.

Sharma offered an explicit departure from Anand Patwardhan's sonic mediations in particular. Referencing Patwardhan's 2012 film *Jai Bhim Comrade*, which conveys and celebrates the songs and protests of Maharashtrian *dalits* (previously 'Untouchables', that is, the most socially stigmatized category subject to caste oppression). She told me, '*Jai Bhim Comrade* was inspiring for many reasons. [...] It made me critically aware that the popular music I was working with [in many ways] did not carry the voice of the oppressed, it did not carry radical politics. It had a political charge but not a charge that I was easily accepting of or in awe of'. She continued, clarifying, 'for me that image of those ten thousand men' amassing on Juhu beach to celebrate an ethnic Hindu identity 'is edgy, and it is not comfortable. After all, the same group can be mobilized to attack the Muslim Other tomorrow' (Sharma 2014).

That is to say, although Bhojpuris in Mumbai are subject to oppression in some ways, they are strong in number, and the Bhojpuri identity that is represented in the auditory, visual and physical phenomenon of the crowd, as well as in the sexual dynamics of their sung performances, involves an aggressive Hindu masculinity. The distancing effect of Sharma's sound design thus resonates with the filmmaker's own ambivalence about that identity and her desire to disrupt for her audience the affective circuits that music produces in these performances. Sharma does not want to convey 'the voice of the people' uncritically, but rather to draw ethnographic attention to its operation. The voice of the film is not, or at least not simply, 'the Bhojpuri voice'.

'The Bhojpuri voice' and identification rather than identity

In her insightful work on Bhojpuri cinema, Kathryn Hardy presents a further complication to the idea of a singular 'voice' in the Bhojpuri case. Charting the production of Bhojpuri cinema, most of it made in Mumbai, and its consumption across migrant networks linking Mumbai and north India, Hardy argues that Bhojpuri subjectivity constitutes not a 'geographically linked category of linguistic belonging', so much as a 'geographically unmoored category of affects' (2015: 148). This insight emerges from Hardy's observation that there is no 'isomorphic dyad of language and geography' that one can identify to constitute Bhojpuria; 'the linguistic limits of Bhojpuri are always in question and the geographic limits of Bhojpuri have not been coterminous with any state' (145). Many producers of Bhojpuri film are speakers of Hindi or Marathi. And many fans of Bhojpuri films don't identify as Bhojpuri speakers (even that identification is difficult, given the indeterminacy of the definition of the Bhojpuri language), but rather identify *with* the experience of mobility and displacement that Bhojpuri films – and especially the songs they contain – tend to express (148).

Hardy's insights are useful for analysing the presence of another key performer whose voice animates the soundtrack of Sharma's film alongside Kallu's and Yadav's. Kalpana Patowary is a female singer who, at least in the film, does not perform to massive crowds of mobile-phone-wielding men, instead recording her performances in studios. Indeed, Patowary is one of the most successful recording artists in the Bhojpuri scene. She identifies with the playful sexuality that Bhojpuri songs express, especially the songs that women sing at weddings that contain double-entendres that do not so much fetishize as express female sexuality (Niranjana 2006). In an interview in *Bidesia*, Patowary asserts, 'I have sung many such songs and I enjoyed them fully. In a sense it was a way to experience a woman's sensuality. This eroticism, I expressed through my songs.' She also identifies with the legal and geographical marginality of the Bhojpuri community; as a Bhojpuri music recording artist, her recording studio, deep in the heart of the Bhojpuri neighbourhood, is demolished along with all surrounding structures in the course of the film. Moreover, Patowary identifies with the migratory experience of Bhojpuri speakers. She is also a migrant, but one from Assam rather than the Bhojpuri homelands of Uttar Pradesh or Bihar.

In the scenes in Patowary's recording studio, the singer constantly pauses and revises her vocal performance in conversation with a producer who remains off-screen. Sharma commented to me about this:

> During the edit it seemed correct that she [Patowary] was constantly repeating, breaking, interrupting her song, allowing for a very different set of conversations to come into play as opposed to the interruptions during the other performances: a truck passing through the audience, people coming up and rewarding money, people throwing stones, a *lathi* charge etc. (Sharma 2014)

Figure 5.4 Kalpana Patowary recording in her studio. Still from *Bidesia in Bambai* (Surabhi Sharma 2013). Courtesy Surabhi Sharma.

The iterative disruptions to Patowary's singing are meant to resonate with what I have referred to as the 'pirate aesthetic' that emerges from the imperfect conditions in which Bhojpuri music practices take place. Here Sharma refers primarily to the imperfect and lively *social* conditions in which Bhojpuri performances occur: the interactivity of their performance genre, the inadequacy of many of their ad hoc performance sites and the adversity of the urban social terrain in which they are trying to claim belonging. To Sharma, at least in this quote, the pauses and repetitions in Patowary's recording sessions are deeply true to the Bhojpuri aesthetic, the same aesthetic that Sharma and her editors sought at times to mimic in their sound design.

However, it is important to note another, perhaps more obvious reason for the many moments in which Patowary pauses her singing, consults with her producer and revises her vocal performance. It is explicit that she often does so in order to correct her pronunciation of Bhojpuri words, or to ask about their significance to Bhojpuri speakers and to comment on their significance to her. These are moments of cultural and linguistic translation and interpretation. They reflect Patowary's marginal relationship to Bhojpuri language and culture, which are themselves complexly correlated with one another, as well as being difficult to define, mobile and marginal.

Patowary's presence and vocal performance in the film prompt us to consider the complexity of the relations between song, community sentiment and identity in the Bhojpuri case. Patowary and, as we know from Hardy's work, many others identify *with* the experience of dislocation and oppression expressed in Bhojpuri music rather than identifying *as* Bhojpuri in any simple way. The idea of 'the voice' as metaphor implies agency, which usually presumes an agent. Here there is no clearly defined agent, or at least no isomorphic relationship between language and geography that would indicate an identifiable 'Bhojpuri' subject who could be said to possess and express a 'Bhojpuri voice'. Instead, in this case, Bhojpuri music facilitates a community of affect with which people identify in indirect, incomplete or multiple ways.

Sharma's film depicts these complex dynamics of sound, sentiment, identity and community, her sound design resonating with them but also serving to contain them for critical consideration. This reflects an approach to 'the voices' – in terms of the perspectives and the vocal practices – of her subjects that departs from the approaches of many Indian documentary filmmakers, particularly those in the documentarian-as-advocate mode that has gained dominance in Indian documentary.

Listening to the voices in and of Indian documentary film

When scholars challenge us to consider whether documentaries speak 'for, alongside, or with' their subjects (see Ruby 1991), they are usually posing important questions about perspective, authorship and authority. In the Indian case, as elsewhere, these are crucial questions. However, I have sought to illustrate here the importance of considering 'the voice' in documentary film as not only a representational phenomenon but also a sonic one. Anthropological approaches to sound incline us to consider 'the voice' as an expressive but also impressionable mediator between the human body and the social. One must attend to the cultural and contextual specificities of the ways in which sound is experienced, practised and understood, as well as the ideologies and practices of personhood, community and communication at play in any particular act of voicing.

All of these considerations are crucial to comprehending the politics of voicing at play not only in everyday life in India, but also in the non-fiction films that seek to document and, often, intervene in that life. Bringing together the sonic and the metaphorical in the analysis of 'the voice' in the case of Surabhi Sharma's work sheds light on the complexities at play in Bhojpuri music, in which the affective efficacy of sound and human vocalization conveys an intimate appeal to multiply displaced people rather than asserting an identity with which listeners simply identify. It also illuminates the complex politics of representation in contemporary Indian documentary, which are being worked out by innovative filmmakers like Sharma in large part in the register of sound.

Notes

1 In this way the voice is similar to the skin as theorized by Turner (2012 [1980]).

2 Double-entendres and sexual themes are typical of certain all-female Bhojpuri musical contexts such as wedding preparation ceremonies (see Niranjana 2006). In the case discussed here, however, female performances of salacious songs help to titillate and celebrate male sexuality.

References

Aitken, I. (2006), *Encyclopedia of Documentary Film*, New York: Routledge.

Appadurai, A. (1993), 'Number in the Colonial Imagination', in C. A. Breckenridge and P. Van Der Veer (eds), *Orientalism and the Postcolonial Predicament*, 314–340, Philadelphia, PA: University of Pennsylvania Press.

Beck, G. (1993), *Sonic Theology: Hinduism and Sacred Sound*, Columbia, SC: University of South Carolina Press.

Bhatti, S. and C. Pinney (2011), 'Optic-Clash: Modes of Visuality in India', in I. Clark-Deces (ed.), *A Companion to the Anthropology of India*, 224–240, Walden, WA: Wiley-Blackwell.

Connor, S. (2001), 'Edison's Teeth: Touching Hearing'. Available online: http://www.stevenconnor.com/edsteeth/ (accessed 20 April 2017).

Daniel, E. V. (1984), *Fluid Signs: Being a Person the Tamil Way*, Berkeley, CA: University of California Press.

Erlmann, V. (2004), 'But What of the Ethnographic Ear? Anthropology, Sound, and the Senses', in V. Erlmann (ed.), *Hearing Cultures: Essays on Sound, Listening and Modernity*, 1–21, New York: Berg.

Faudree, P. (2012), 'Music, Language and Texts: Sound and Semiotic Ethnography', *Annual Review of Anthropology*, 41: 519–536.

Feld, S. and D. Brenneis (2004), 'Doing Anthropology in Sound', *American Ethnologist*, 31 (4): 461–474.

Feld, S., A. A. Fox, T. Porcello, and D. Samuels (2005), 'Vocal Anthropology: From the Music of Language to the Language of Song', in A. Duranti (ed.), *A Companion to Linguistic Anthropology*, 321–345, Walden, MA: Blackwell Publishing.

Fernandes, N. (2012), 'Colloquial Music Reflects Changes in Rural Indian Romance', *New York Times India Blogs*. Available online: india.blogs.nytimes.com/2012/06/20/colloquial-music-spawns-a-culture-of-romance/ (accessed25 March 2015).

Gade, A. (2004), *Perfection Makes Practice: Learning, Emotion and the Recited Quran in Indonesia*, Honolulu, HI: University of Hawaii Press.

Hansen, T. B. (1999), *The Saffron Wave: Democracy and Hindu Nationalism in Modern India*, Princeton, NJ: Princeton University Press.

Hardy, K. (2015), 'Constituting a Diffuse Region: Cartographies of Mass-Mediated Bhojpuri Belonging', *BioScope: South Asian Screen Studies* 6 (2): 145–164.

Hirschkind, C. (2006), *The Ethical Soundscape: Cassette Sermons and Islamic Counterpublics*, New York: Columbia University Press.

Keil, C. and S. Feld (1994), *Music Grooves: Essays and Dialogues*, Chicago, IL: University of Chicago Press.

Kothari, R. (1989), *State against Democracy: In Search of Humane Governance*, New York: New Horizon Press.

Labelle, B. (2010), *Acoustic Territories: Sound, Culture and Everyday Life*, New York: Continuum.

Larkin, B. (2004), 'Degraded Images, Distorted Sounds: Nigerian Video and the Infrastructure of Piracy', *Public Culture*, 16 (2): 289–314.

MacDougall, D. (2006), *The Corporeal Image: Film, Ethnography and the Senses*, Princeton, NJ: Princeton University Press.

Manuel, P. (1993), *Cassette Culture: Popular Music and Technology in North India*, Chicago, IL: University of Chicago Press.

Marriott, M. (1990), *India through Hindu Categories*, Delhi: Sage.

Matzner, D. (2014), '*Jai Bhim Comrade* and the Politics of Sound in Urban Indian Visual Culture', *Visual Anthropology Review*, 30 (2): 127–138.

Nichols, B. (1994), *Blurred Boundaries: Questions of Meaning in Contemporary Culture*, Bloomington, IN: Indiana University Press.

Niranjana, T. (2006), *Mobilizing India: Women, Music, and Migration between India and Trinidad*, Durham, NC: Duke University Press.

Pinney, C. (1997), *Camera Indica: The Social Life of Indian Photographs*, Chicago, IL: University of Chicago Press.

Revill, G. (2000), 'Music and the Politics of Sound: Nationalism and Auditory Space', *Environment and Planning*, 18: 597–613.

Roy, S. (2002), 'Moving Pictures: The Postcolonial State and Visual Representations of India', *Contributions to Indian Sociology*, 1&2: 233–263.

Ruby, J. (1991), 'Speaking For, Speaking About, Speaking With, or Speaking Alongside: An Anthropological and Documentary Dilemma', *Visual Anthropology Review*, 7(2): 51–66.

Samuels, D. W., L. Meintjes, A. M. Ochoa, and T. Porcello (2010), 'Soundscapes: Toward a Sounded Anthropology', *Annual Review of Anthropology*, 39: 329–345.

Sharma, S. (2007), Interview with author (13 June).

Sharma, S. (2014), Email to author (15 July).

Spivak, G. (1988), 'Can the Subaltern Speak?' in C. Nelson and L. Grossberg (eds), *Marxism and the Interpretation of Culture*, 66–111, London: Macmillan.

Sterne, J. (2003), *The Audible Past: Cultural Origins of Sound Reproduction*, Durham, NC: Duke University Press.

Stoller, P. (1989), *The Taste of Ethnographic Things: The Senses in Anthropology*, Philadelphia, PA: University of Pennsylvania Press.

Sundaram, R. (2015), 'Post-Postcolonial Sensory Infrastructure', *E-Flux Journal*, 63 (April). Available online: http://www.e-flux.com/journal/64/60858/post-postcolonial-sensory-infrastructure/(accessed 19 April 2017).

Taussig, M. (1993), *Mimesis and Alterity: A Particular History of the Senses*, New York: Routledge.

Taylor, W. (2002), 'Penetrating Gazes: The Poetics of Sight and Visual Display in Popular Indian Cinema', *Contributions to Indian Sociology*, 36: 297.

Trawick, M. (1990), *Notes on Love in a Tamil Family*, Berkeley, CA: University of California Press.

Turner, T. S. (2012 [1980]), 'The Social Skin', *Hau: Journal of Ethnographic Theory*, 2 (2): 486–504.

Vohra, P. (2011), 'Dotting the 'I': The Politics of Self-less-ness in Indian Documentary Practice', *South Asian Popular Culture*, 9 (1): 43–52.

Weidman, A. (2014), 'Anthropology and Voice', *Annual Review of Anthropology*, 43: 37–51.

Wolf, N. (2002), 'Make It Real: Documentary and Other Cinematic Experiments by Women in India', PhD diss., Europa-Universitat Viandrina, Frankfurt/Oder.

6

Hum, Buzz, Gurgle: Ecological Soundscapes in Poetic Ecodocs

Shilyh Warren

This chapter explores the construction of voice, and in particular, the production of non-human ecological soundscapes, in poetic documentaries about the challenges of the Anthropocene. Offered as a new field of acoustic studies within ecology, 'soundscape ecology' refers to the study of all sounds that emanate 'from a given landscape to create unique acoustical patterns' (Pijanowski et al. 2011: 203). Unlike prior studies of acoustics in the natural sciences, soundscape ecology considers the unique sonic collages of specific sites 'that are biological, geological, or anthropogenic' (Pijanowski et al. 2011: 204). Three recent, short, poetic films with complex ecological soundscapes – Canadian Peter Mettler's *Petropolis* (2010), American Erin Espelie's *Silent Springs* (2011) and the British media artist Roz Mortimer's *Invisible* (2006) – offer sonic and affective alternatives to the anthropocentric voice of crisis that dominates both conventional ecodocs and conversations in the media about the health and future of our planet's ecosystems. Following political philosopher Jane Bennett, I argue that these films, in large measure through their non-traditional ecological soundscapes and non-human voices, produce experiences of wonder and enchantment rather than fear and pessimism. The alternative voices in these films – emanating from animals, water and machines, rather than human experts – aggregate connective tissue between what Bennett calls ethical thoughts and ethical behaviours.

According to Paula Willoquet-Maricondi (2010), recent scholarship in eco-cinema has tended to distinguish between three modes of ecocritical film practices: fictional feature films like *Erin Brokovich* (2000) and *Promised Land* (2012), expository documentaries like *An Inconvenient Truth* (2006), *Crude Awakening* (2006) and *Gasland* (2010) and poetic, experimental documentaries such as *Fog Line* (1970) and *Riverglass* (1997). With little variation, the consensus around these films goes as follows: fictional features arguably raise awareness but create no space or need for action, because, as in most conventional feature films, a capable protagonist heroically saves us all. As spectators, we easily abrogate our ethical and political responsibilities to the larger-than-life activist on screen. Expository documentaries on the other hand deal with the problem in a way that links knowledge to action. Either these documentaries include an overt call to action or they implicitly assume that by raising cognitive awareness, they create a catalyst for political engagement. Helen Hughes (2014: 12) observes that even though environmental psychologists have known since at least the 1970s that 'knowledge and awareness' are never enough to generate new ethical behaviours, most environmental advocacy organizations continue using communication strategies that emphasize knowledge as a sure path to action.

Poetic documentaries have been said, however, to engage less in a call to action than in the reorganization of perception: to experience time and nature in a new way is to potentially develop a new ethic towards the environment. Scott MacDonald (2013: 19) explains that these films provide 'visual/auditory training in appreciating the experience of an immersion within natural processes'. For MacDonald, 'The job of an ecocinema is to provide *new kinds of film experience* that demonstrate an alternative to conventional media spectatorship and help to nurture a more environmentally progressive mindset' (2013: 20, emphasis in original). But as David Ingram (2013: 47) argues, MacDonald's optimism about the impact of experimental ecofilms may depend overly on 'wishful thinking'. Challenging MacDonald's emphasis on techniques such as the long take and slow pacing of experimental films, Ingram asserts the value of entertainment films comes from their cognitive and affective accessibility, which he also finds valuable for ecopolitics. If MacDonald and Ingram represent two poles of thinking about the value of experimental ecodocs, the two scholars also represent a common disregard for the production and function of voice in documentary.

In contrast, I argue that the construction of voice in some poetic documentaries about nature offers an alternative to these two silent positions. That is, while I agree with MacDonald that poetic documentaries encourage new modes of perception and affective attunement between nature, cinema and spectators, focusing on voice demonstrates how such films offer meaningful – though admittedly ephemeral – moments of enchantment that potentially link ethical thinking to ethical behaviour. Similarly, while I agree with Ingram that a dose of wishful thinking and highbrow assumptions influence MacDonald's desire to link cinematic strategies like long takes and environmental awareness, I also continue to insist that non-optical aesthetics play a neglected role in furthering ethical behaviours towards the environment. Furthermore, some expository documentaries, because they so often rely on a panicked, terrifying and potentially paralysing anthropocentric voice of crisis potentially hamper the relationship between thinking and acting, which as Hughes reminds us, is so often assumed to be a natural consequence of knowledge and awareness. Thus, this chapter proposes that the affective imperative of the non-human voice presents a worthwhile alternative mode of ethical encouragement in experimental ecodocs.

In *Green Documentary*, Hughes identifies three major responses to environmental concerns evident in contemporary documentaries, which in her estimation always combine epistemological and affective dimensions: contemplative, ironic and argumentative. Hughes emphasizes tone, as in the tenor that attaches to the information conveyed. Among the three tones, the contemplative tone, manifest in poetic, narrative films like *Manufactured Landscapes* (Jennifer Baichwal 2006) and *Sweetgrass* (Lucien Castaing-Taylor and Ilisa Barbash 2009), focuses most intensely on the complexity of environmental consciousness, allowing the viewer the time and space 'to hold in mind and contemplate the interconnections between knowledge and emotions about the environment' (2014: 13). However, Hughes's notion of tone addresses primarily the affective dimension of information transmission rather than the quality of voice. Like, Hughes then, I seek a more robust understanding of the connections between the epistemological and affective aims of environmental documentaries. However, my notion of voice in documentary pressures the anthropocentric assumptions undergirding Hughes's theory of communication and tone. Certainly, environmental consciousness is generated by a complex network of knowledge, feeling and context, which, for Hughes, documentaries offer in a range of tones. However, I seek to specifically explore the possibilities for environmental consciousness and activism that arise when the human voice is decentred. I focus here on the interplay of human, geological and animal voices within ecodoc soundscapes, arguing that it is crucial to the affective possibilities the films generate.

Crisis

Stuart Hall has argued that the cultural production of crises, particularly since the 1970s, evidences similar 'underlying features' and is oriented in a similar 'direction of travel' (2011a: 705). For Hall, 'crises are moments of potential change' with no guaranteed result, although they share three likely outcomes: society may move on 'to another version of the same', 'to a somewhat transformed version' or towards radical transformation (2011b: 37–38). As we know from history, radical transformation is the least likely outcome. More often than not, crisis discourse serves the state's impulse to preserve the status quo or consolidate its capital and ideological interests. The production of an environmental *crisis* then, in the realm of documentary, risks appropriation by forces that would maintain rather than revolutionize the status quo. Crisis talk is thus potentially complicit with discourses that precisely endanger the environment. Thus, the trend in environmental documentaries to frame or communicate arguments about the Anthropocene in terms of crisis risks paralysing rather than catalysing action.

Crisis ecodocs dominant today, including films such as *An Inconvenient Truth, The Eleventh Hour* (2007), *Food, Inc.* (2008), *The Cove* (2009) and *Gasland* (2010), share many formal and thematic elements. They also share a quality of voice; that is, they depend on the human voice (as well as the male human body) to frame and express the problem in anthropomorphic and (phallo)logocentric terms. In these films, it is often the voices of men either on or off-screen that passionately urge us to understand the impact of corporate greed and exploitation on the environment. Picture Leonardo di Caprio in *The Eleventh Hour* on a literal mountaintop, his voice imploring, his hand gestures punctuating his speech: 'The evidence is now clear. Industrial civilization has caused *irreparable* damage. Our political and corporate leaders have *consistently* ignored the *overwhelming* evidence.' Another male voice off-screen says, 'Not only is it the eleventh hour; it is eleven fifty-nine'. In both cases the words convey the information and the tone of voice, as Hughes would call it, is emphatic, anxious, pleading. The message is clear: it is time to panic.

This approach to environmental crisis, I argue, is incommensurate with the magnitude of the environmental problems we face in the twenty-first century. Bill Nichols asks how documentaries can 'conjure or restore for the viewer those orders of magnitude appropriate to the full dimensionality of the world in which we live' (1991: 230). He asks how documentary films offer the possibility of a subjective experience of the world that is viscerally meaningful and to a degree commensurate with the magnitude of the referent itself. How do these crisis documentaries make the earth felt? Conventionally voiced crisis documentaries tend to reduce the experience of a radically compromised planetary ecosystem to a feeling of alarm and fear; the magnitude of the earth as problem outweighs the magnitude of the earth as home. Retreat or recoil from the entreaty to feel afraid would be a logical response to the emphatic tone of voice in these documentaries, despite the fact that many of these films conclude with attempts at semi-hopefulness and encouragement (Turn off your lights! Buy an electric car! Bike to work!).

In a slightly different context, political philosopher Jane Bennett also worries that stories about the world in a state of crisis 'do more to undermine ecological concern than raise awareness of ecological destruction' (2001: 91). Indeed, Bennett's main objective in *The Enchantment of Modern Life* is to counter the characterization of contemporary modernity as necessarily disenchanted; that is, lacking and alienating, technologized and consumed, grinding and regrettable. In what she calls the narrative of disenchantment, neither nature nor culture can be sites of wonder, surprise or joy; and the self that is configured by disenchantment primarily experiences loss and cynicism. Disenchantment, Bennett explains, makes it hard to say yes to the world, to a loving life, to generosity. She asks, 'What's to love about an alienated existence on a dead planet?' (2001: 4)

Whereas disenchantment occasionally points to the premodern or the divine as a source of enchantment, Bennett is convinced 'enchantment-power' resides in complex networks of entities, material and natural, secular and ordinary. 'Enchantment', Bennett explains, comes with a 'mobilizing rush' – an affect of possibility – that can inspire or motivate the energy required to transform ethical thinking into ethical practice. In other words, the affect of enchantment, 'with nature but also with commodities and other cultural products', Bennett argues – in her next book, *Vibrant Matter* – assembles a vital link to ethical behaviour (2010: xi).

My aim here is to mobilize Bennett's theory of enchantment to challenge the assumption, forwarded by Ingram, for example, that poetic documentaries have limited potential – that is the idea, championed by MacDonald, that poetic documentaries primarily offer states of altered perception and thought. The non-human voices of *Petropolis, Silent Springs* and *Invisible* provoke the possibility of enchantment, which is to say the possibility of an affective encounter valuable to the activist endeavour of ethical practice. If 'sound has an integral role in shaping the affective contours of our day-to-day lives' (Thompson and Biddle,

2013: 11), then the ecological and largely non-human soundscapes generated by poetic documentary shape a valuable relationship between ethical feelings and environmental activism.

Petropolis (2010)

Petropolis bears the functional subtitle, 'Aerial Perspectives on the Alberta Tar Sands'. As this description suggests, Mettler and his team constructed the documentary entirely from aerial footage taken from a helicopter, thousands of feet above the area in question. Commissioned by Greenpeace of Canada, the documentary clearly hopes to transform viewers' literal, optical understanding of the Alberta Tar Sands, and the natural and material consequences of the bitumen extraction process necessary to the fabrication of North American petroleum. For the first five minutes of the film, the camera glides above a pristine landscape, granting a privileged panorama of Edenic proportions. The sweeping views of Alberta's boreal forest and rivers are accompanied by harmonic sounds of natural life: rushing water, chirping frogs, calling birds, percussion that mimics the beating of a heart; and also by complementary technological sounds: digitized beeps and clicks, metallic clangs and dings, at one point the ebb and flow of radio static. Twelve textual overlays narrate key details about the scale and history of the landscape, the nature of its most valuable resource: bitumen, and the dire consequences of its extraction. That is, in just under six minutes, the images, text and sounds combine in a complex network of audiovisuality that conveys exploitation, pollution and irreparable damage as well as the material presence of ecological wonder – all without a human voice.

Thus, when the rhythm of mechanized beeps, clicks and tones increases in speed and volume and the frame gracefully lifts to reveal the proximity and scale of the destruction of the landscape (as in Figure 6.1), the impact is heartbreaking – that is, while refineries are expected in this film, the sudden shift within a single take from natural beauty to the machinery and exhaust of petroleum production is disorienting. Adjacent to the greenery clearly exists a monumental shattering of nature, which is conveyed in both image and sound. There are no trees to see and no frogs to hear in the soundtrack, only synthetic monotones carried on too long and helicopter blades spinning threateningly. The bulk of the documentary stays here where the tar sands have been transformed by extraction and fabrication processes. Graceful long takes capture flaming

Figure 6.1 A toxic lake in *Petropolis* (Peter Mettler 2010) shatters the idyllic scenery of the opening of the film. Greenpeace/Eamon Mac Mahon.

exhaust towers, rushing gallons of polluted water, construction vehicles loading and hauling, miniature figures padding across bitumen fields and occasionally, patches of green and diminutive houses.

As this description demonstrates, the documentary provides a limited amount of linguistic information and no vocalized argument or human mouthpiece. After the twelve textual overlays only images and non-verbal sounds drive the affective force of the documentary, and yet these aggregate into a dense data system – a network of audiovisual information that aims to terrify and amaze, or in Bennett's terms, surprise and enchant. Certainly it is possible to read the documentary as evidence of planetary disenchantment, for example when the text indicates that only 3 per cent of the possible action in the forests has yet transpired. But Mettler's film disturbs this web of pessimism and hopelessness. Rather, the documentary is expansive, highlighting actual destruction as well as the presence of natural life, thriving there still. By focusing on trees and rivers, the film offers visual reminders of the longevity of the Holocene even as it stresses the challenges of the Anthropocene; sounds from nature often exist in harmony with artificially produced sounds of technological processes, suggesting the possibility of coexistence rather than dissonance or retreat. The problem, therefore, is not that technology exists; after all, the helicopter is what makes possible this new apprehension of the planet, this new perspective on the wounded landscape. The problem, as the ecological soundscape conveys, is that the scale of the exploitation has gone unchecked. What future is possible in this story about the Anthropocene and its voracious hunger for fossil fuels?

Mettler's response to this question begins with the sudden introduction of the human voice, and a return to time before the birth of the Anthropocene. Thirty-six minutes into the film Mettler's human voice enters the soundscape for the first time. The camera glides above the tailing ponds, improbably beautiful as sunlight shimmers over polluted waters. Mettler's voice-over narrative begins in the year 1783 with the story of the first aerial flight: the 2-mile, 8-minute hot air balloon ride taken by a sheep, a rooster and a duck and witnessed by 130,000 spectators, including Marie Antoinette, Louis VXI and the French court. Mettler connects his own petroleum-fuelled voyage over the tar sands with this maiden aerial voyage, telling us that we are once again faced with a 'new perspective of a landscape we cannot comprehend from the ground'. This is what the documentary asks us to do: not to comprehend an archive of 'information' about the tar sands, but rather to apprehend a new perspective on the effects of this extraction process on the planet. His final spoken words are: 'What will we do next?'

After Mettler's brief monologue, we visually return to the trees and to the technical–natural soundscape. In the frame, a telling path snakes through the forest. This time the camera moves through space with speed, and a steady rhythm, a kind of march. There are tyre marks below; there is likely a human plan for this area. Is the futurity of the planet already predicted by the Anthropocene, or is another future possible? Despite its stark visualization of the Anthropocene, *Petropolis* draws the spectator in, not with an anthropocentric and logocentric appeal to human empathy or comprehension, but rather towards a sensuous journey, one that is both heartbreaking and heart holding. The powerful entreaty voiced by a network of human and non-human sounds in *Petropolis* suggests that the appeals of traditionally voiced expository documentaries may not constitute the ideal path to ethical behaviour. Rather, both concern and enchantment here speak in a range of voices, natural, technological and human; they speak to attachment rather than crisis. *Petropolis* apprehends the crisis of the Anthropocene in a comprehensive way, viscerally and emotionally, without resorting to dominant androcentric optical and aural logics of crisis.

Silent Springs (2011)

Erin Espelie's film, *Silent Springs*, in contrast, makes more extensive use of the human voice, but harmonizes human language with the sound and rhythm of water. The short film takes its name and its cue from Rachel Carson's landmark 1962 exposé of the damage caused by industrial pesticides. Carson lamented the toxic future heralded by chemical pollution: the 'silence' implied by her title. Yet, Espelie's film is precisely remarkable for its attention to sound, and the relationship between human vocality and nature's sonority. Espelie's implied subjects are frogs, but these are clearly allegorical figures, creatures on the front lines of the industrial agricultural pollution of rivers and lakes; the thin-skinned creatures that evidence the harmful effects of pesticides since they lack alternative defences or habitats. Like Mettler, Espelie draws our attention to dimensions of scale by offering an alternative audiovisual perspective. If the damage at stake in the Alberta Tar Sands demands a macro view, then the chemical pollution of the planet's waters demands a consideration of the microscopic, the damage wrought to the planet's non-human species, cell by cell (see Figure 6.2).

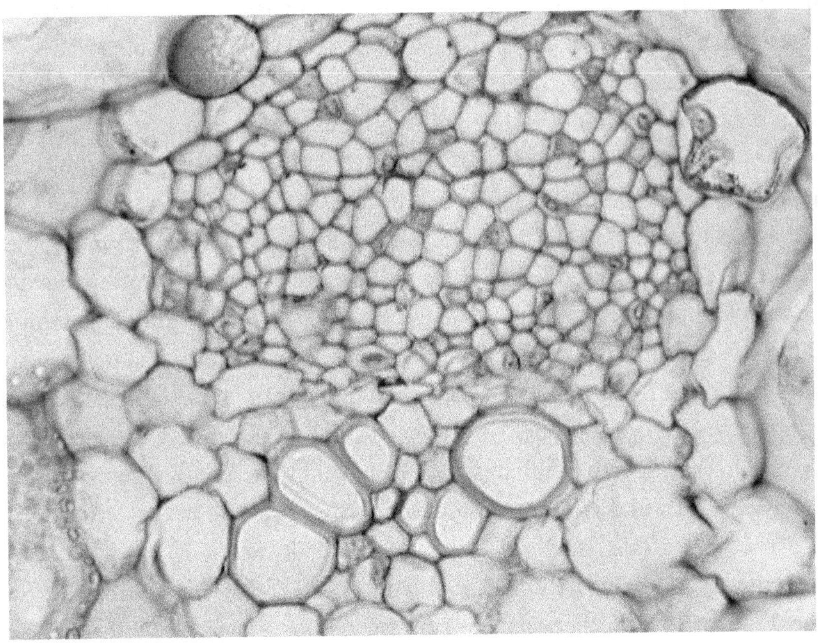

Figure 6.2 At the centre of *Silent Springs* (Erin Espelie 2011), a montage of cellular material in microscopic detail. Courtesy Erin Espelie.

Espelie's film opens with close-ups of human objects: an antique microscope, ageing slides, a distressed wooden table – subtle reminders of the dawn of the Industrial Revolution. In the film's opening sequence, we see the microscope in close-up and we look through its viewfinder at the mounted slides. We hear the camera's lens working, zooming in and out, attempting to focus on the microscope, which allows for a focus on the plates; we hear the scrape of the glass plate sliding against the metal housing. In this first sequence, the film shifts unpredictably, moving through a range of unlikely angles on the microscope and glimpses of names on the slides: Silkworm, Silverberry, fish belly flesh. A few minutes into the film, a man's voice speaks about optical technology and the challenges of seeing and staging the specimens, but Espelie wants us to hear these passages from nineteenth-century monographs on optics in more general terms. Indeed, sound and image relations throughout the film foreground the difficulty of seeing, focusing, working. 'It may be impossible', 'insuperable', 'difficult', the voice tells us, to work with the scientific tools. But it also may be 'impossible to discontinue', he says towards the end of the film.

Silent Springs thus reflexively thematises the act of audiovisual apprehension. By framing this theme within nineteenth-century discourse about scientific discovery, Espelie also connects these shifts to the dawn of the Anthropocene. Throughout the film, we see and hear images of living contemporary frogs in crisp digital images as well as grainy mid-century frogs from prior educational documentaries. We also see slides of frog blood, and at the centre of the film, an elegant, enchanting montage of stills of cellular material. The film thus offers sequences that replicate the naturalized scale of human hearing and vision, but also ask us to come to terms with the technology by which we apprehend these images: the camera and the microscope. The use of archival footage and sound emphasize the range and age of the technologies of seeing, drawing attention to the fact that the audiovisual data and knowledge have been readily available to us for decades. The aim of *Silent Springs* thus is not to create new and more precise knowledge about the Anthropocene, but to assemble a question about the required perspective: What does it take to fully and bodily apprehend the damage? What will this apprehension achieve?

Silent Springs also includes several long static takes of the Villa d'Este fountains near Rome, Italy. The fountains, a UNESCO World Heritage Site, signal a surprising pact between humans, technology and nature. In the sixteenth century when the Villa was built, human ingenuity and aesthetic commitments resulted in technological transformations of nature that remain beautiful, purposeful and enchanting. In the present, shots of the fountains in *Silent Springs* provide a soothing counterpoint to the malformed frogs and the scientific revelations about pesticides and pollution. Water is thus both the film's theme (since it is the frogs' habitat) and its aesthetic method of affective calibration for the viewer. Indeed, *Silent Springs* borrows its rhythm and affective energy from the constantly flowing water. *Silent Springs* proceeds with more activist urgency than Mettler's film, and emphasizes a sober, scientific problem, but by layering discourses about optics and science over time, as well as a panoply of images, sounds and technological methods of apprehension throughout centuries of human presence, *Silent Springs* also creates space for connectivity and wonder.

Both *Silent Springs* and *Petropolis* eschew the idea of a pure and untouched kind of nature, the pristine and divine nature that Bennett says exists only in the 'elsewhere' in the narrative of disenchantment. Both films construct historical perspective about the challenges and rewards of the Anthropocene. Through aural means, the two films also suggest that the aim of environmental documentaries should not be only to provide new information, but also to assert a new means of connection – a new frequency for attunement – with the problems that we already know exist. In their use of complex soundscapes that blend human, non-human and technological voices, both films also acknowledge instances when the interconnectedness of nature, culture and technology has been positive and important for humans in the Anthropocene. However, they refuse the facile suggestion that human ingenuity and technology are obvious solutions to the problems humans have created. Neither of these films offers casual, androcentric calls of alarm, nor simple melodies of hope, but together, they decentre the human voice to cultivate an experience of connection and attachment between humans, animals, machines and the environment. Although the film repeats the phrase, 'it may be impossible', *Silent Springs* also urges us to feel the sonic caress, elemental certainty and timelessness of water.

Invisible (2006)

Roz Mortimer's *Invisible* presents a related, but also distinct range of questions, since the film relies partially on the bodies and voices of Inuit women to decentre the habitual male voice of ecodocs and achieve its affective impact. Although it treads into the uneven terrain of the ethnographic documentary desire to know the indigenous other, *Invisible* also seeks to captivate rather than alarm and achieves this in part through alternative sound and image relations. According to Mortimer's voice-over in the film, quite by accident, scientists have discovered that the breast milk of Inuit women in the northern-most regions of Canada contains the highest levels of chemical contaminants found on Earth: seven times higher than women in the rest of the world. I say 'quite by accident' because scientists sought out Inuit women assuming that they would find a control group – pure specimens against which to compare more obvious candidates in developed urban centres with higher pollution. The irony of this assumption and the tragedy of its discovery are the crux of the story in Roz Mortimer's *Invisible*.

Although Inuit communities of Baffin Island in Canada continue to hunt and eat traditional foods, such as seal, narwhal and polar bears, they are now aware that these marine mammals harbour in their fatty tissue dangerous levels of chemical pollutants. For childbearing women, this scientific data is especially concerning. The communal act of sharing raw and frozen meat has been tainted by the knowledge that by consuming these foods women introduce unknown quantities of chemicals into their bodies and the bodies of their biological children. We learn from the film that technological advances have made chemical compounds so minute that pollutants can actually permeate the placenta, affecting the unborn foetus in ways we are only just beginning to understand. Furthermore, the only way for the body to expel these chemicals is through breast milk. Women's bodies thus become the conduits for pollution to enter the bodies of their children. *Invisible* asks viewers to experience the dilemmas Inuit communities now face. How will Inuit women renegotiate the relationships between the sea and the kitchen table, between their traditional, cultural practices and their bodies, and between their babies and their milk? As one woman tells us in *Invisible*, 'we have a lot to lose … an entire way of life'.

By this summary alone, it is clear that *Invisible* shares a good deal of information with the spectator – about the recent history of chemical development and use, about the cultural practices of the Inuit and about the Arctic Sea and its changing ecology. Most of this scientific and anthropological knowledge is conveyed through traditional documentary conventions, such as interviews, voice-over and text. Mortimer's own voice opens the film, and two major expert testimonies surface and complement each other in the soundtrack. Textual overlays often reiterate the direst facts and figures. And importantly, *Invisible* includes filmed interviews with two Inuit mothers, who poignantly convey their experiences with pregnancy and breastfeeding, and the struggle to incorporate new scientific data into a traditional way of life.

Feminist documentaries characteristically depend on women's voices and women's bodies in filmed interviews and voice-overs to facilitate identification between viewers and filmed subjects – strategies that emerged in the highly politicized period of women's filmmaking in the 1970s. That Mortimer depends on these strategies to create new epistemologies and new modes of documentary affectivity connects her work to a vital canon of feminist documentaries, and indeed highlights the vitality of feminist perspectives on the environment and the gendered disparities of ecological ruin. This contradicts the impulses of most canonical ecodocs, which construct a sense of urgency and a call to action by foregrounding the human voice and the human body, particularly in male voice-overs and with cinematography that replicates human vision. Certainly humans are the cause of the planet's troubles; however, poetic ecodocs that challenge androcentric and masculinist logics of voice and vision suggest that dominant films may not be the shrewdest response to the challenges of the Anthropocene.

The interview segments with women in *Invisible* give life and specificity to the larger calamity of chemical pollution and climate change. They also construct an alternative epistemology about chemical pollution, based in personal, gendered and ethnic experience – complementing the scientific evidence and data, which the documentary also includes. These are the ways the film attempts to construct and convey knowledge about chemical pollutants in our environment and in our bodies. And while knowledge shapes ethical thinking, even crisis-docs demonstrate that spectators need something more – more embodied, more affective, more felt – in order to move from thinking ethically to behaving ethically. In *Invisible,* Mortimer experiments with text, sound, animation and voice-over as she weaves the affective fabric of the documentary. The film balances its attention to ecological calamity and ethnic and gender disparity with performative sequences, which focus on women's bodies and their voices, and also mobilize curiosity, enchantment and wonder.

Surfacing throughout the film and traversing the surface of the Arctic is an enigmatic feminine figure in a bizarre red dress (see Figure 6.3). When she first appears in the opening of the film, she cuts a silent, surprising path through the snow and ice. What is she doing here? What is she wearing? What does a woman in a hooded red dress have to do with chemical pollution, rising sea temperatures and the demise of an entire way of life? Indeed, what does she have to do with science, knowledge and environmental protection?

Figure 6.3 Traversing the surface of the Arctic in *Invisible* (Roz Mortimer 2006) is an enigmatic feminine figure in a bizarre red dress. Courtesy Roz Mortimer.

The red dress symbolizes what the bodies and blood of the animals conceal: harmful chemical compounds that would slip into and among the healthy vascular networks of mammalian bodies. She embodies thus the invisible, silent spring that looms over these communities. And the anachronistic style of her dress makes subtle connections to the history of colonialism – crucial to the history of the documentary form as well as contemporary neo-colonial enterprises that pollute places on earth where the inhabitants are responsible for so little of it.

But this body also literally sings of possibility – the possibility of an alternative relationship to knowledge, and maybe the possibility of an alternative ecological future. The film's soundscape features a unique musical score, including a theremin composition, free-yoik from Sami musician Wimme Saari and live operatic throat singing from globally recognized Inuit performer Tanya Tagaq. Costumed variously in her colonial red dress and then in a traditional amaut parka, Tagaq performs several renditions of throat singing. We learn from the text that these traditional throat songs of Inuit women derive from the sounds of the animals in the ecosystem: whales, seals and walruses. Thus, the musical soundscape of the documentary reveals a harmonious contract between humans, animals and the environment, which stands in stark contrast to the dominant habits of consumption and exploitation that characterize late capitalism.

Mortimer (2014) hoped *Invisible* would challenge 'western ideas of knowledge, science and authority', and convey the sense that the day to day lives of the Inuit 'involve/d constant analysis of the environment that was bound up with their myths and traditions, and a visual attunement to the weather and the land – and I wanted to incorporate something of this in how I presented the environment and the crisis within the film'.[1] These twin goals emerge most clearly in another scene, when the two singers hold reams of paper in front of them. As the Arctic wind blows, sheet-by-sheet, the papers, which represent Western knowledge and scientific data, disappear into the snow and ice of the landscape. These performative sequences accrue in the documentary, assembling vitality without emergency, and the possibility for ethical movement without fear or paralysis in crisis. In my reading, these performative scenes inspire what Bennett calls moments of enchantment, which she defines as 'acute sensory activity', both pleasurable and disruptive. To focus on the affective dimension of the documentary is to home in on the film's ethical desires. In *Invisible,* ethical behaviour towards the environment requires an affective encounter with a new epistemological frame, defined sonically and visually outside of the androcentric norm.

In different ways, *Petropolis, Silent Springs* and *Invisible* all make vital links between humans, technology and the environment, especially in their soundscapes and the ways they produce alternative voices. Whereas *Petropolis* and *Silent Springs* consider the environmental hazards of the Anthropocene with sound and image relations that de-emphasize the logics of the human body, sight and voice, *Invisible* emphasizes the materiality of the body, especially those of indigenous women. As a result, *Invisible* invites critique about its underlying ethnocentrism, which serpents throughout the history of documentary filmmaking, especially about native peoples. However, Mortimer's *Invisible* offers an alternative to the legacy of 'taxidermic' ethnographies, in Fatimah Tobing Rony's (1996: 100) terms; that is, the film, though made by a white, European filmmaker offers an ethical representation of the Inuit communities on Baffin Island and grants its indigenous subjects vocal and sonic authority. Mortimer foregrounds a *local* analysis of the problem by interviewing Inuit scholars as well as local community members, and by foregrounding the singing of Inuit performers. Moreover, the film stresses the interconnectedness of vast regions of the globe, rather than emphasizing the isolation or inherent difference of Inuit communities. It is not their 'difference' that the film explores, but rather their local, situated expertise about the chemical predicament that threatens their communities, and will eventually threaten ours, too. Therefore, the performance of throat songs in the documentary should enchant us not merely because the songs are unique and culturally specific, but because they represent an agreement between humans, animals and the environment – a contract to behave ethically towards one another. *Invisible* finds valuable balance between dominant androcentric, scientific epistemologies and alternative and indigenous modes of knowing and feeling. The film is urgent without using scare tactics, convincing without crisis. By appealing broadly to the spectator as sensorium, *Invisible* inspires as many questions as answers, and not just: How can we know? How can I tell you? But also: What can we do?

Conclusion

In this chapter I have argued that because they rely on androcentric vocalized narratives of crisis, most environmental documentaries reiterate what Bennett calls the narrative of disenchantment, which impedes ethical action. What Bennett misses is that every narrative is uttered by a specific voice, and the voice itself carries symbolic and aesthetic possibilities and limits. Experimental, issue-driven documentaries like *Petropolis, Silent Springs* and *Invisible* activate alternative soundscapes, complicate the assumed authority of the androcentric and masculine voices of conventional documentaries and thus catalyse the possibility of enchantment, which is to say the possibility of an affective encounter valuable to ethical practice. While we concede that the Anthropocene has made us all vulnerable, these films stress the disproportionate vulnerability of some more than others, and at the same time validate the primacy and complexity of the voice as key to the aesthetic experience of inspiration – a vital element of activist practice.

Note

1 According to Mortimer (2014), 'Interestingly some (western) scientists have found these more affective parts of the film a bit challenging.'

References

Bennett, J. (2001), *The Enchantment of Modern Life*, Princeton, NJ: Princeton University Press.

Bennett, J. (2010), *Vibrant Matter*, Durham, NC: Duke University Press.

Hall, S. (2011a), 'The Neoliberal Revolution', *Cultural Studies*, 25 (6): 705–728.

Hall, S. (2011b), 'The March of the Neoliberals', *The Guardian* (12/9). Available online: http://www.theguardian.com/politics/2011/sep/12/march-of-the-neoliberals (accessed 1 November 2016).

Hughes, H. (2014), *Green Documentary: Environmental Documentary in the 21st Century*, Bristol and Chicago, IL: Intellect.

Ingram, D. (2013), 'The Aesthetics and Ethics of Ecofilm Criticism', in S. Rust, S. Monani, and S. Cubitt (eds), *Ecocinema Theory and Practice*, 43–62, New York: Routledge.

MacDonald, S. (2013 [2004]), 'The Ecocinema Experience', in S. Rust, S. Monani, and S. Cubitt (eds), *Ecocinema Theory and Practice*, 17–42, New York: Routledge.

Mortimer, R. (2014), Email correspondence with the author.

Nichols, B. (1991), *Representing Reality*, Bloomington, IN: Indiana University Press.

Pijanowski, B. C. et al. (2011), 'Soundscape Ecology: The Science of Sound in the Landscape', *BioScience*, 61 (3): 203–216.

Rony, F. T. (1996), *The Third Eye: Race, Cinema, and Ethnographic Spectacle*, Durham, NC: Duke University Press.

Thompson, M. and I. Biddle (2013), *Sound, Music, Affect: Theorizing Sonic Experience*, New York and London: Bloomsbury.

Willoquet-Maracondi, P., ed. (2010), *Framing the World: Explorations in Ecocriticism and Film*, Charlottesville, NY and London: University of Virginia Press.

7

Literally Documenting Her Voice: Valie Export's Vocal-Visual Experiments with Screen Media

Helen Hughes

Valie Export is an artist, filmmaker, curator and educator whose work has been central to the feminist avant-garde.[1] Beginning in the 1960s the radical,[2] politicized approach taken by women artists working both independently and in collectives in many different countries came to be associated with the Second Wave of feminism. A central achievement of the movement has been to destabilize the understanding of art as the product of male genius (Schor 2016: 23). In the process of building gender consciousness, the strengthening of the female voice within and across the institutions of politics, education and culture has also been a goal. As with the image, however, the very idea of the female voice has required deconstruction.

This questioning of the voice, and its role as the carrier of patriarchal language, features strongly in Valie Export's oeuvre. In an early work from 1968, a *sehtext* (see text) entitled *fingergedicht* (finger poem), the performer uses sign language to say 'I speak the sign with signs showing speech' (1980: 104), which uses the body to make speech visible while also enacting Julia Kristeva's theory of the speaking subject who makes or unmakes herself (1989: 295). Export's voice is now the mature voice of an artist who has been productive since the 1960s and has engaged with the complex problems of asserting female identity across many different media. The very establishment of her artistic practice through the name VALIE EXPORT as combined name and logo, only to be written in block capitals,[3] is a relic and a demonstration of the role of media and language in the construction of her voice and identity. Export's work is experimental in that it creates images and sounds in a process of performance that tests ideas about the voice by embodying them. It is also appropriately considered as documentary in that the performances are recorded and then manipulated and inserted into a social document.[4]

Stella Bruzzi has pointed out that feminist theorists have exploited a significant cultural difference in the reading of the male and the female voice in the traditional documentary voice-over that means the male voice is used to assert universality while the female voice is the voice of specificity. She argues that this gives female commentary the power to expose 'the untenability of documentary's belief in its capacity for imparting "generalized truths" faithfully and unproblematically' (Bruzzi 2000: 59). The female voice thus has the capacity to act as the 'rebellious voice' in Export's terms, but the disruption that it brings about is more profound than merely offering an alternative to male universality. The loss of generalized truth involves not merely an alternative gendered voice, excluding the 'voice of god', or even allowing only 'literal' or diversified voices, but rather provokes a rejection of the idea that a voice has a unified and fixed identity capable of expressing the 'truth' of the subject that speaks.

A sequence of works begun in 2007 – a performance documentation, an installation and a film – has brought Export's work on the female voice and identity into focus once again. The three works, *the voice as performance act and body* (2007), *glottis* (2007) and *I turn over the pictures of my voice in my head* (2008), use a laryngoscope to display and record Export's voice as she reads. Each one is centred in a startling way on the synchronized visual and acoustic documentation of the voice through the visualization of vocal cords, the glottis and the section of the larynx that surrounds them. This is an unusual experiment with the idea of the 'audiovisual contract' (Chion 1994: 1–123). The strangled delivery of the text does not maintain the pure clarity of 'voco- and verbocentric' sound in film that Michel Chion describes as the ultimate goal of synchronization (1994: 6), but it does 'add value' to the mobile shapes of the glottis, by structuring the spectator's vision of it as a living vehicle in the chain of vocal expression.

Chion criticizes an experiment the essay filmmaker Chris Marker made with voice-over in *Letter to Siberia* (1957). He describes Marker's synchronization of three different texts with the same sequence as 'the simple situation of a political opinion slapped onto images' (1994: 7), a gesture which, Chion argues, fails to demonstrate how sound structures image. In Export's three experiments the synchronization looks horrifyingly real, but as an abject image–sound combination that is only just about rendered tolerable through the association between the text and the medical context it originates from. Her work shares with Marker's film a strategy of laying bare the audiovisual contract through its flaws and its literalness. Reciting texts continuously and – due to the position of the laryngoscope – with audible difficulty at moments of swallowing or during articulations that close the vocal tract, Export's glottis is in constant motion. Reinforcing the fascination with the image–sound combination, the text she reads maintains its focus on the subject of the voice, the vocalizations themselves competing with the drama of the image and with the shifting shapes of the glottis. An essay on the voice in its own right, Export's work examines the nature of the relationship between voice and identity. More specifically, her opening analysis of the voice as 'sign', 'sign of images', 'sign of sensuality' and 'sign of the symbols' expresses her understanding of a fundamental split between the external voice and an inner acoustic voice she later turns to.

The video *The Voice as Performance, Act and Body* (2007) documents a performance that took place at the 52nd Venice Biennale in 2007. Valie Export was represented by a number of works under the title *Der Schmerz der Utopie/The Pain of Utopia*. The title of the performance video maps out its meanings, beginning with the idea of the 'voice as', a characteristic move for Export. Taken from Wittgenstein, this 'voice as' notion understands vocal signification as a process of resemblance rather than as a medium that posits a relationship with truth. The voice as performance, as act and as body is visually inscribed in the video, which shows the artist sitting at a desk reading out texts. With her is Erhard Suess, the medical doctor who is credited for the 'camera laryngoscope', along with Robert Stockinger. Stockinger removes each page as it is read, but he is present because the procedure requires the support of a further person for safety. The video documents the participants in the laryngoscopy as well as the audience for it; behind the performers, four screens display images of the larynx moving simultaneously as the texts are read. No ultimate truth regarding Export's voice is revealed: instead the performance and video elaborate the sonic, somatic and social contexts for the voice's production.

The voice of documentary, in Bill Nichols's sense of a 'moiré-like' textual authorial construct (Nichols 1983: 18),[5] can be found in the video's simultaneous display of the medical and medial structure through which the speaking voice comes to attach itself to the moving image of the glottis. Although the performance is captured largely in a simple tableau, there are also sequences of close-up footage to capture the insertion and removal of the laryngoscope. Valie Export's editing of the performance video also includes cuts from the tableau to the laryngoscope images of the glottis, a close-up in a different format, constructing a more literal movement into the somatic apparatus of the voice.

In his book *Audio-Vision* Michel Chion also gives an account of 'causal listening' in which he reminds the reader that 'a sound often has not just one source but at least two, three, even more' (Chion 1994: 27–28). Export's video *The Voice as Performance, Act and Body* represents several places for the voice. Valie Export herself is most obviously the source as she sits and reads. The camera records a close-up of her face showing where the tube of the laryngoscope enters via her nose. In these shots her lips move, placing the voice at her mouth, which Chion has labelled the 'symbolic place of vocal production' (1999: 127). Behind her, however, the image of the glottis is visible on large screens, moving in a way that seems to synchronize with the sound of the voice. The tube held by the doctor is evidence that these images are connected to the voice that is heard. *The Voice as Performance, Act and Body* shows that these images have been split between four screens, demonstrating the capacity of audiovisual media to relocate the voice, multiply it and then also install it into several objects, including the human body, at once.

Another object visible on screen is the microphone, which is understood as part of a chain of mechanisms translating the voice into a digitally recordable form which can be translated back into sound and amplified. The microphone records the vocal sounds as they are heard outside the body – the voice as socially produced – whereas the laryngoscope records the medically produced, internal voice. The continuous soundtrack is synchronized with both the external and internal shots (the latter made with the laryngoscope) and thus demonstrates that the voice is located in multiple sites. The overall effect runs counter to Chion's model where location and synchronization produce meaning, as the images and sounds dissipate rather than secure signification.

On her website entry for the performance video, Export has published the opening text from the film entitled *The Rebellious Voice, the Split Voice*. In it she labels the voice as 'suture', 'seam', 'cut' and 'tear', denying that the voice is 'body or spirit', 'language or image'. She also labels this deconstructed voice, which seems to serve as both the space between and the stich uniting textual bodies, as 'my identity', perhaps associating the words with her training as a young woman in the Design Department of the Textile Industry Institute for Teaching and Research in Vienna (Export, 2017).[6] In characterizing the voice as a medium (a suture or stitch joining textiles), Export asserts the social role of the voice through the embodied action of the speaker, and thus retains the strictly constructivist understanding of the individual voice that is typical for postmodern feminist activism. The split voice functions as a suture, an activity depending not on established connections but on a continuous process of speaking. Women are thus enabled by the rebellious voice in their efforts to break down gendered power structures. This understanding is, however, utopian in the way suggested by the exhibition title *The Pain of Utopia*. To imagine a better world, to imagine gender equality, is to let go of the safety of existing identities and to recognize the inability of the voice to guarantee a stable, singular identity. It also characterizes voices as plural and potentially in competition with each other.

Valie Export's individual experience with the tools of medical practice is, of course, not unique. In *Postopera*, in which she gives an account of the 're-invention of the voice – body relationship' in contemporary post-dramatic and postmodern opera (Novak 2016: 6), Jelena Novak describes seeing a video projection of her vocal cords while speaking during a rhinoscopy examination (2016: 153). Rather than closing the gap between the voice and the body, Novak notes, the mismatch became more profound: 'I could not perceive any apparent connection between movements of my interior organs and the expressions of my voice' (153). Her conclusion about the examination and the insight it brought for her was confirmation that 'the key to understanding the body-voice relationship does not lie in the empirical, but in a symbolic connection between the two' (153).

The synchronization of the glottis with the voice in Export's performance work challenges Novak's conclusion in that Export shares this experience of disconnection but then asserts that her voice is indispensable to her. Her reflection on the voice as performance, act and body turns into a search for understanding her dependence on her joy and pain as her voice comes and goes. In the process of reflection her voice appears to become separated from her but this separation prompts a scream which reveals another inner voice, which she describes as 'the sea in me'.

The reference to the sea is significant as it is a self-quotation from a poem she placed at the start of her book *Körpersplitter* (Export 1980). In this poem she writes that the sea has become too big for her and the verbal poem ends when she strikes out her tongue and turns to visual thought. At the end of the poem is a colon and the rest of the volume consists of experimentation with images until she returns to painting and poetry and the *sehgedicht* mentioned above. *The Voice as Performance, Act and Body* reconnects with this history of a personal struggle with the voice and the subsequent turn to the language of images and gesture. Just as she maintains the constructed nature of the visual body sewn together from internal and external perspectives, so too she demonstrates the process of the construction of the voice from multiple joints and spaces: 'My voice is the trace of my individual body as it is of my social body. It sews the parts, the sewing patterns of my ephemeral identities together.'

A statement of a similar kind is made with *Glottis*. After the performance at the Venice Biennale in 2007, the space in which the performance took place was filled with an installation work consisting of the four screens of the images of the glottis. In this version each screen had its own soundtrack which is started out of sync with the others. In Valie Export's online catalogue a poem appears that defines the vocal cords as 'symbols of the voice'. Export explores the point of division:

> The glottis, the vocal cords, are symbols of the voice
> They divide two phenomena
> The voice inside, the breath
> And the voice outside
> The phenomenon of vocalization
> Of speech formation. (Export 2007)

It is worth pausing a moment to analyse this poem formally as well as thematically. Formally, the poem obscures the moment of the bifurcation of the voice through the placement of words: the inside voice is breath and the outside voice is concerned with speech formation. After the statement that asserts the symbolic nature of the glottis and vocal cords a thesis is introduced about the organs marking a division. This division is spatially and temporally defined, which is reflected in the positions of words in the poem. The words 'inside', 'breath' and 'outside' are in a spatial sequence which also reflects the process of speaking. 'Breath' comes after the 'voice inside', separated with a comma, but then 'the breath' is also separated from 'the voice outside' with a line break – a bigger break in spatial terms. The 'phenomenon of vocalization' then appears as a unified concept so that although the glottis is a symbol marking a point of division, it is still part of the physical process rather than detached from it. It seems paradoxical to assert that such a precisely verbally and visually defined division is symbolic, but this reveals the power of the media – visual, acoustic and graphic. The glottis has become symbolic as soon as it entered the systems of representation, through participating in the breath and the voice outside.

Export's work has brought the video laryngoscope as a medium into the context of her body of work and hence into the world of the feminist avant-garde. 'The Rebellious Voice, the Split Voice' points to the history of her own feminist art practice (see Export 1989). Her work as a whole goes beyond the idea that the image of the glottis deepens the gap between the voice and the body. Rather she has expanded the number of points along the vocal tract, and along the production of meaning, that can be claimed as symbols of the voice. Each point splits the body in the flow of time and movement and then extends it out into the world through the prosthetic apparatus. The glottis is thus used to symbolize the point at which the interiorization of cultures of perception is exteriorized but is itself only a place where air enters the vocal tract and begins to vibrate.

The interplay between the internal and external body has been important for Export's work over a long period. Taking its cue from a quotation from R. D. Laing's *The Divided Self* in Export's film *Syntagma*, Kaja Silverman's essay 'Speak, Body' explores how Export has used the split screen to examine the relationship between the 'specular' body and the 'sensational' body (Silverman 2000).[7] In her reading of the film, Silverman argues, 'Export first thematizes the sensational body as a kind of corporealized speech, and later metaphorizes the duality of the body as a text read in two different languages' (Silverman: 2000). Because the text by Laing is read in two languages – German and English – and then later in the film the texts and images appear through different media such as photography, books, films and video, Silverman concludes that Export's technique is pointing to a parallel between the body of representational media and the body of the woman: 'The image is both video and book at the same time, just as the body is both image and sensational ensemble' (Silverman 2000: 19). Writers on Export's films have disagreed about the meaning of this parallel. Silverman reads it as an assertion of the dual nature of the body when she says, 'the body is neither one nor two; it is two in one' (Silverman 2000: 23), while acknowledging that Roswitha Mueller reads Export's work as 'about the signifying power of the body, not about the embodiment of the signifier' (1994: 198).

The full dimensions of this disagreement are illustrated by considering the number of screens and their arrangement in the installation *Glottis*. Although it is common practice for video installation artists to use four screens – they form a cross or a square or a tower – Export's choice has a resonance in the context of debate about female identity and its exclusion within a patriarchal society. Export arranges the four slightly concave monitors around the corner of the room in a semicircle, which places the installation into the context of extended debates about two Greek terms used to represent the essential nature of the female as receptacle: *hystera* or womb and *Korē* or undefined space. In *Revolution and Poetic Language* Julia Kristeva uses the term *Chora* to refer to the vocal tract, 'to denote an essentially mobile and extremely provisional articulation constituted by movements and their ephemeral stasis' (Kristeva 1985: 25). In her work *Speculum: The Other Woman*, Luce Irigaray analyses both terms and their identification with the female alongside an analysis of the mirror as the visual apparatus that signifies the self's recognition of itself. All three terms come together in her deconstruction of Plato's Allegory of the Cave in the *Republic* which, she argues, reveals a crucial step in the exclusion of the female from Western philosophy (Irigaray 1985a).

Export's work *Glottis*, based as it is on the insertion of an instrument, a 'speculum', to look inside the body, can be understood as entering into a visual experiment with the tools of medical examination to search for the missing female voice. As such it is also a reflection on her own engagement as a visual artist with the problem of the female subject and an analysis of her fellow feminist artists' struggles in 'The Real and Its Double' (Export 1988–89) where she cites Irigaray's second work *This Sex Which Is Not One* (Irigaray 1985b), a reflection on *Speculum*. The Allegory of the Cave, recounted by Socrates, is meant to explain the existence of 'ideal forms' behind visible objects. These perfect forms explain the origins of the existing imperfect and varying shadows of human experience. Irigaray deconstructs the scene, however, and argues that the Cave in fact represents a masculine denial of material origin in the womb or *hystera* and a replacement of it with signs or language: 'The Idea is intended to be real, independent of the art of mirrors. And yet the world from end to end is organized as *mimēsis*; re-semblance is the law' (Irigaray 1985a: 149–150).

Glottis consists of four DVD screens that project images of the vocal cords and the space in between – the glottis – in a round frame and synchronized with the voice of the artist. They are not synchronized with each other: the shapes move continuously in different ways and the voices are a cacophony of statements. The Cave then is here transformed and deconstructed again by Export into another kind of space that reflects on the body as it is turned inside out in Irigaray's analysis. It is an attempt at creating a space within mimetic space which allows for difference and for an understanding of both the image and the voice as signs. Important for this demonstration of mediation and multiplication is the choice of four monitors as they can accumulate in pairs, avoiding the reduction to the single entity which reproduces itself as 'Sameness' (Irigaray 1985a: 361). Irigaray argues that it is the desire for one single origin that excludes the possibility of two sexes, male and female, and indeed of difference more generally. In her latest book Irigaray focuses on the breath as the 'elusive' point of origin of the individual that is in fact 'ecstatic' or outside ourselves and the correct target for attempts at transcendence (Irigaray 2017: vii).

A discussion of the third related work, the short film *I Turn over the Pictures of My Voice in My Head*, which edits together the sound of the voice and a single rectangular image of the glottis synchronized into a linear sequence, provides an opportunity to think through the idea that the synchronization of the voice with the glottis rather than the lips represents the possibility of a female voice in a world that allows both male and female and hence difference. In the case of this film the shift from the mouth to the larynx as the symbolic (and embodied) location of the voice is complete. A number of paradoxes emerge from this process of 'nailing down' the voice as Marguerite Duras once described the process of synchronization: 'Duras coined the idea that the contemporary cinema stringently requires voices to be *nailed down* to bodies. It's this nailing, which is for her a form of cheating, that she tried to break with in *India Song*' (Chion 1999: 130). As the voice is firmly attached to the interior body with no alternative offered, the idea that the voice is the final expression of identity and guarantor of existence appears stronger. This idea emerges in Samuel Beckett's existential play *Not I* in which the character *Mouth*, represented on stage and on screen only by the mouth, is unable to negate existence as long as she speaks. But Export challenges the ideas of Duras and Beckett by speaking of a celebration as speech comes and goes: 'I await the return of my voice. Whenever it comes back in the moments of gestures as echo of mirrored skin, as fragments of yearnings, as freedom from compulsions, I don't really have to understand its return, but it is indispensible to me.' At this point Export quotes from Lucien Israël, a psychoanalyst and writer on the history of hysteria. His 1976 book *L'hystérique, le sexe et le médecin* (lit. The Hysteric, Sex and the Doctor) was translated into German in 1983 as *Die unerhörte Botschaft der Hysteria* (lit. The Unheard of [as in outrageous] Message of Hysteria). Israël is a double-edged reference. Although he extends the idea of hysteria from its construction as a female malady to male patients, he also understands feminism as a form of successful hysteria. As Elaine Showalter has pointed out, 'female activism becomes merely a constructive pathology, and feminism only a healthier form of hysteria. It does not occur to Israel to label Flaubert or Sartre a successful hysteric' (Showalter

1995: 290). Export's reference nevertheless picks up the idea of the hysteric's scream as a positive rather than a negative force and celebrates it for its receptive and combative qualities.

Despite acknowledging the need to break down the patriarchal image of woman, the feminist avant-garde has also engaged in an intense search for an authentic female voice. Hélène Cixous's explosive essay 'Laugh of the Medusa' is a call to women to find 'l'écriture feminine' inside themselves and has been criticized as the expression of an essentialist understanding of womanhood (Cixous 1976). The essay begins, however, not with a definition of what women's writing is but with 'what it will do' – and becomes an extended exhortation asking women who feel called to write to be courageous enough to write in a way that expresses their experience as specifically that of women in women's bodies. The essay is then in itself a demonstration, an act of self-realization in which Cixous herself claims her own writing as a means through which she can kick-start her own history as well as the history of women. She proceeds to argue her sense of what it means to write her body, which means: 'to break up, to destroy; and to foresee the unforeseeable, to project' (Cixous 1976: 875). Export has repeatedly expressed her understanding of the media as a means to change women's social reality. Already in her manifesto written to accompany *MAGNA: Feminism Art and Creativity*, an exhibition of women's art, in 1972 she wrote in the context of a growing feminist debate about the relevance of art she argued 'let women speak so that they can find themselves':

> The arts can be understood as our medium of self-definition adding new values to the arts. these values transmitted via the cultural sign-process, will alter reality towards an accommodation of female needs. (Export 1973)

The feminist avant-garde was an art movement that rarely allowed the visual or the verbal gesture to stand alone, combining both in constant process of deconstruction. Carolee Schneemann's *Interior Scroll* (1975), in which she pulled a text from her vagina and read it out, was an iconic performance that looked like a rebellion against an internalized patriarchy. It was indeed a reaction to a philosophical critique of her work by a fellow feminist. One of many performances designed to reconnect women with their bodies, this was also one of many to turn to the visual to make the connection between female language, female sexuality and the interior body.

The documentation of performance has been an integral part of the feminist avant-garde. Works such as *Interior Scroll* or Export's famous action

Tapp- und Tastkino (Tap and Touch Cinema) (1968) were disseminated through still and moving photography, published in books or shown at collective art shows. One of the most profound aspects of the documentation process has been the way in which the media – principally photography, film and video – were not only a means to observe and record ephemeral performances and action pieces spontaneously carried out in public spaces, but also used to capture private performances with an intimate or sometimes no audience present at all (Schor 2016: 30). This radical practice of independent art making eventually merged with auto-ethnography described as 'AvantDoc' by Scott MacDonald (MacDonald 2015) and is exemplary for the way in which it demonstrated the new kind of art practice necessary to establish a space for excluded voices, one in which the speaking subject makes or unmakes herself, to paraphrase Kristeva again (1989: 295).[8]

The voice that is projected in feminist art in the 1970s and that is recalled in *I Turn over the Pictures of My Voice in My Head* is thus the product of a long process in which the artist is concerned with reinventing the voice and its connection with bodies and images in a gender conscious way. In the film Valie Export synchronizes her voice uncompromisingly with the image of her glottis, creating a composite object that is both a literal voice and a voice-over that reflects on itself. The authority of the commentary is under constant scrutiny in the effort to attach the movement of the glottis and the larynx with the sounds of speech.

But consistent throughout the monologue is the idea of the voice as separate, something akin to a partner in the struggle for self-realization or consciousness. Although 'the breath of life is its source', there is the complaint that 'it distances itself from me, it goes its own way, it betrays me'.[9] The sense of the voice as itself a partner in a dialogue is reinforced by the use of quotation. Export names Mladen Dolar (with difficulty as the laryngoscope hinders her capacity for articulation) as she quotes a passage from his book *A Voice and Nothing More* (Dolar 2006). She does not just repeat the text word for word but rather modifies it into a personalized version. While Dolar writes:

> We constantly inhabit the universe of voices, we are continuously bombarded by voices, we have to make our daily way through a jungle of voices, and we have to use all kinds of machetes and compasses so as not to get lost. There are the voices of other people, the voices of music, the voices of media, our own voice intermingled with the lot. (2006: 13)

Export says:

> I have to make my way daily through a jungle of voices. I have to use all kinds of compasses and knives so as not to get stuck there. These are the voices of other people, the voices of animals, the voices of civilization, of music, of the cities, the voices of the heights and the depths, the voices of war, the voices of the media and in the midst is my own voice. The voice I own (I turn over the pictures of my voice in my head, 2008) (my translation).

The adaptation of Dolar transforms the quotation from an illustration into an interrogation and appropriation of ideas through the action or performance. The quotation from the book is treated itself as a medium and is partially accepted but also subverted by the incorporating media of performance art and digital video. In an interview at the end of the 1990s as she was experimenting with the boundaries between analogue and digital media, Export explained her process: 'Every medium has its own relation to what we call reality and can be subverted in its own specific way to counteract this effect' (Mueller 1994: 212). She described her media works as 'medial anagrams': 'What interests me is that even minimal shifts in context will bring out differences in signification for the same unit of representation. It is a kind of language system for image production in the technological media' (Mueller 1994: 213).

The testing of Dolar's ideas, and also Dolar's response to the film in a lecture after she sent it to him (Dolar 2008), is, in terms of Export's work, part of a continuing experiment with other voices in and around her work. She thus engages in an audiovisual version of Irigaray's insistence that to reintroduce the female is also to destroy the unity of patriarchal philosophy and to maintain the flow of difference, the continual exchange of the breath.

At the end of her discourse she quotes Mladen Dolar again in claiming 'the voice is the instrument, the vehicle, the medium, the meaning, the goal' only to describe how the images elude her attempts to fix them as 'pictures' of her voice:

> I turn over the pictures of the voice in my mind. I shape them into circles, squares, pyramids. I turn them, tilt them, but I cannot fix these pictures of my voice in my mind's eye.

The performance documentation, *The Voice as Performance, Act and Body*, the DVD installation *Glottis* and the short film *I Turn over the Pictures of the Voice in My Head* all project the voice as something more than utterance, more than the performance of speech. Taking the glottis as symbolic of division, these works document and develop the process of creating the meaning of voice as a process of dialogue, mediation and synchronization.

Notes

1 In defending the idea of a specifically feminist avant-garde Schor writes 'the feminist art movement's historic and pioneering achievements in the art of the past four decades is not in dispute. The protagonists of the feminist avant-garde wrote manifestos and pamphlets, established numerous women artists' associations and journals, articulated a critique of art institutions, organized their own exhibitions, created groundbreaking work in terms of form as well as content and sought to fuse art with life. In short their activities manifest all the characteristics of the avant-garde predominantly associated with male artists' (Schor 2016, p. 18). She also notes: 'Remarkably, many of these artists pursued similar strategies without knowing each other's works' (21).

2 The first 'gender conscious' work is Carolee Schneemann's *Eye Body* performed and documented through photography in 1963 (Stiles 1998: 297).

3 In art historical texts such as the catalogue accompanying the exhibition *The Feminist Avant-Garde* the name is generally written in block capitals. In her essay on *Feminist Actionism* Export's name appears in more conventional orthography, her academic name perhaps diverging from her artistic one (Export 1989).

4 A historical account of the cat-and-mouse relationship between feminist avant-garde and feminist documentary film and video can be found in the introduction to *Feminism and Documentary* (Waldman and Walker 1999).

5 'By "voice" I mean something narrower than style: that which conveys to us a sense of a text's social point of view, of how it is speaking to us and how it is organizing the materials it is presenting to us. In this sense "voice" is not restricted to any one code or feature such as dialogue or spoken commentary. Voice is perhaps akin to that intangible, moiré-like pattern formed by the unique interaction of all a film's codes, and it applies to all modes of documentary' (Nichols 1983: 18).

6 The textile industry features often in feminist filmmaking. See, for example, Sue Clayton and Jonathan Curling's *The Song of the Shirt* (1979) discussed in *Towards Other Cinemas* (Eshun 2017).

7 Export quotes almost word for word from Laing's *The Divided Self: An Existential Study in Sanity and Madness* (Laing 1960): 'The body clearly occupies an ambiguous transitional position between "me" and the world. It is, on the one hand, the core and centre of my world, and on the other, it is an object in the world of the others' (131).

8 Writing about feminist performance video of the eighties and nineties Alexandra Juhasz writes: 'You could call the vast majority of this video work "documentary" as it is composed primarily of images of a videomaker's unscripted performance as she breaks rules of female propriety. Recorded on tape for later exhibition these are documents of a politicized (usually autobiographical) self-expression, a woman

performing and archiving her defiance against the rules of sex and gender' (Juhasz 1999: 96).

9 The gender of the word for voice in German in feminine so that when she refers to her voice in the third person, as she does throughout the performance, Export addresses it as 'sie' or she.

References

Bruzzi, S. (2000), *New Documentary: A Critical Introduction*, London and New York: Routledge.

Chion, M. (1994), *Audio-Vision: Sound on Screen*, New York, Chichester and West Sussex: Columbia University Press.

Chion, M. (1999), *The Voice in Cinema*, ed. and trans. C. Gorbman, New York: Columbia University Press.

Cixous, H. (1976), 'The Laugh of the Medusa', *Journal of Women in Culture and Society*, 1(4), 875–893.

Dolar, M. (2006), *A Voice and Nothing More*, Cambridge, MA: MIT Press.

Dolar, M. (2008), *Mladen Dolar, What's in a Voice (Transkript)*. Available online: http://philosovereign.blogspot.co.uk/2014/02/mladen-dolar-whats-in-voice-transkript.html (accessed 3 January 2017).

Eshun, K. (2017), 'On Her Devolves the Labour: The Cinematic Timetravel of *The Song of the Shirt*', in Clayton, S. and L. Mulvey (eds), *Other Cinemas: Politics, Culture and Experimental Film in the 1970s*, x–xx, London: I.B.Tauris.

Export, V. (1973), 'Women's Art: A Manifesto (March 1972)', *Neues Forum*, 228: 47.

Export, V. (1980), *Körpersplitter. 1, Konfigurationen: Fotografien 1968–77*, Linz: Edition Neue Texte.

Export, V. (1988–89), 'The Real and Its Double: The Body', *Discourse*, 11(1): 3–27.

Export, V. (1989), 'Aspects of Feminist Actionism', *New German Critique*, 47: 69–92.

Export, V. (2007), *Der Schmerz der Utopie/The Pain of Utopia*. [Art] (Venice Biennale). *Glottis* (2007) [DVD Installation].

Export (2017), *VALIE EXPORT*. Available online: http://www.valieexport.at/de/biografie (accessed 30 August 2017).

Irigaray, L. (1985a), *Speculum of the Other Woman*, Ithaca, NY: Cornell University Press.

Irigaray, L. (1985b), *This Sex Which Is Not One*, Ithaca, NY and New York: Cornell University Press.

Irigaray, L. (2017), *To Be Born*, Cham, Switzerland: Palgrave Macmillan.

Israël, L. (1976), *L'hystérique, le sexe et le médicin*, Paris: Masson

Israël, L. (1983), *Die unerhörte Botschaft der Hysterie*, trans. P. Müller and P. Pasch, Munich, Basel: Ernst Reinhardt Verlag.

I Turn over the Pictures of My Voice in My Head (2008), [Performance film] Valie Export

Juhasz, A. (1999), 'Bad Girls Come and Go, but a Lying Girl Can Never Be Fenced In', in D. Waldman, and J. Walker (eds), *Feminism and Documentary*, 95–115, Minneapolis, MN and London: University of Minnesota Press.

Kristeva, J. (1985), *Revolution in Poetic Language*, trans. M. Waller, New York: Columbia University Press.

Kristeva, J. (1989), *Language: The Unknown: An Initiation into Linguistics*, New York: Columbia University Press.

Laing, R. (1960), *The Divided Self: An Existential Study in Sanity and Madness*, Harmondsworth: Penguin.

Letter from Siberia (1957), [Film] Dir. Chris Marker, France: Argos Films; Procinex.

MacDonald, S. (2015), *Avant-Doc: Intersections of Documentary and Avant-Garde Cinema*, Oxford and New York: Oxford University Press.

Mueller, R. (1994), *Valie Export: Fragments of the Imagination*, Bloomington, IN: Indiana University Press.

Nichols, B. (1983), 'The Voice of Documentary', *Film Quarterly*, 36 (3): 17–30.

Novak, J. (2016), *Postopera: Reinventing the Voice-Body*, London and New York: Routledge.

Schor, G., ed. (2015), *Feminist Avant-Garde: Art from the 1970s*, Munich and London: Prestel.

Schor, G. (2016), 'The Feminist Avant-Garde: A Radical Revaluation of Values', in G. Schor (ed.), *Feminist Avant-Garde: Art of the 1970s*, 17–71, Munich, London and New York: Prestel.

Showalter, E. (1995), 'Hysteria, Feminism, and Gender', in S. L. Gilman (ed.), *Hysteria beyond Freud*, 286–344, Berkley and Los Angeles, CA: University of California Press.

Silverman, K. (2000), 'Speak, Body', *Discourse*, 22 (2): 8–24.

Stiles, K. (1998), 'Uncorrupted Joy: International Art Actions', in R. Koshalek (ed.), *Out of Actions between Performance and the Object 1949–1979*, 227–329, New York: Thames and Hudson.

The Song of the Shirt (1979), [Film] Dirs. Sue Clayton, Jonathan Curling, UK: Film & History Project, BFI Production Board.

The Voice as Performance, Act and Body (2007), [Performance documentation] Export, V.

Waldman, D. and J. Walker (1999), *Feminism and Documentary*, Minneapolis, MN and London: University of Minnesota Press.

Section Three

Gender, Sexuality and Species

8

The Voice of Mockumentary [1]

Sarah Kessler

Although she may seem to speak in her own voice, and thus provide an unfiltered first-person perspective, the documentary subject rather resembles a ventriloquist's dummy. This is Bill Nichols's conclusion in his defining essay, 'The Voice of Documentary', where he contends that documentary film's frequent emphasis on the voice's subjective authenticity can have the unfunny effect of transforming the documentary subject into a 'puppet' (1983: 27). Manipulated by the filmmaker and the apparatus more generally – since her strings are pulled by the whole kit and caboodle – her voice always, to some extent, speaks in the service of the inaudible but no less ideological 'voice of the text' itself (Nichols 1983: 20). Indeed, as Trinh T. Minh-ha has argued, 'No matter how plural and diverse the voices featured [in a given documentary film], one always has to point back to the apparatus and the site from which these voices are brought out and constructed' (1992: 169). Suffice it to say, both documentarians and documentary film theorists have long been hip to the (documentary) voice's ventriloquism: its speaking from elsewhere, from somewhere other than the body that appears to produce it – its 'brought out and constructed'-ness, to ventriloquize Trinh.[2]

It is this ventriloquism endemic to the documentary form – which relies as much on the voice's evidentiary power as it does on 'visible evidence'[3] – that mockumentary film and television texts most cuttingly mock.[4] After all, puppets are supposed to be funny (right?), so mockumentary gives us a puppet show, a ventriloquist act we can laugh at, and whose 'political stakes … remain relatively low' given the fictionality of its characters (Middleton 2014: 28). But what's so funny about mockumentary's voices? How do mockumentary films and TV shows quite literally 'make fun of' the voices in and of the documentary forms they echo? And what does the voice of mockumentary – which I maintain, following Jason Middleton, has already been 'incorporated [back in] to the language of mainstream documentary film' (2014: 10) – have to tell us about reigning ideas of 'the voice' itself? Through two brief case studies, this chapter argues that while mockumentary texts often directly parody, and in the process draw critical attention to, the documentary subject's ventriloquism, they also, and often simultaneously, proffer this same subject's voice as an authenticating entity. The voice of mockumentary may be funnier than the voice of documentary, but both voices continue to assert powerful claims on reality,[5] albeit in different tones.

Popularized in the mid-1980s by Rob Reiner and Christopher Guest's cult film, *This Is Spinal Tap* (1984), and recently theorized by media scholars Jane Roscoe and Craig Hight, Alexandra Juhasz and Jesse Lerner, and Gerd Bayer, among others, mockumentary films and TV programmes from Woody Allen's *Zelig* (1983) to *Modern Family* (ABC 2009–present) hilariously ape the mechanicity of documentary's talking heads, the over-earnestness or disingenuousness of its speaking witnesses and the self-importance of its disembodied voice-overs, which are regularly coded as white, male and upper middle class. But while such mockumentary film and television texts surely caricature and recontextualize documentary's vocal conventions to foreground the untrustworthiness, subjectivity and manipulation of the voices they project, they also endow these voices with varying degrees of truth-telling power. Mockumentary film and TV texts frequently ridicule their fictional subjects – and through them the voice-giving pretensions of documentary form – by exaggerating the disjunction between the content of a given character's speech and the sound of her voice, such that 'what she says' is directly and amusingly undercut by 'how she says it' (to invoke a useful colloquialism). Mockumentary's tendency to treat the voice not as a mere medium for meaningful language, but as a 'material entity' with sensed attributes whose perception may invoke yet other associations (Schlichter and Eidsheim 2014), thus productively upends documentary's tendency to collapse the capacious category of 'voice' into linguistic testimony.[6] Yet the side-splitting materiality of the mockumentary voice – which in my view (inspired by the emergent transdisciplinary field of voice studies) encompasses such audible vocal properties as accent and intonation (commonly the domain of psycho- and sociolinguistics and phonetics) as well as pitch and timbre (commonly the province of musicology)[7] – may also, I argue, end up replacing the linguistic voice in its typical role as the guarantor of a seemingly fixed, authentic subjectivity.

Because this reversal, whereby the voice's sonic materiality audibly trumps, or 'tells on', its linguistic content, becomes particularly palpable when (homo) sexuality is the butt of a given mockumentary's joke, this chapter focuses on humorous depictions of so-called gay voices in two widely known mockumentary texts: Guest's acclaimed film, *Waiting for Guffman* (1997), and *Modern Family*, a long-running TV series soon to enter its tenth season. Employing what Middleton calls an 'awkward aesthetic' (2014: 23–51),[8] Guest's film uses direct-address interviews and observational documentary strategies to highlight the cringe-inducing feeling that often attends to regular folks speaking 'naturally' before a camera. In the process, the film foregrounds the voice's uncanny ability to 'give someone away' no matter what they might say – an effect demonstrated by *Waiting for Guffman*'s protagonist, whose declarations of heterosexuality are comically belied by his stereotypically 'gay voice'.[9] The everyday ventriloquism parodied in Guest's discomfiting work is also a target of *Modern Family*'s televisual mockumentary, which mobilizes the contemporary reality TV strategy of interspersing direct-address interviews with observational footage to chronicle the puppet show that is the modern family. But while each family member's speech is ultimately affirmed as self-aware regardless of his or her accent, interlopers may become victims of their own material voices – marked as, say, 'gay' without recourse to direct-address.

'Talk like a normal person': *Waiting for Guffman* (1997)

Voice serves as a comic giveaway in Guest's *Waiting for Guffman*, a touchstone of the mockumentary film format and a pioneering example of the production of what Middleton calls 'awkward aesthetics' (2014: 23–51). Per Middleton's formulation, *Waiting for Guffman*'s aesthetic of awkwardness – and the crux of its humour – derives not only from the film's focus on patently '*awkward people*' utterly lacking in self-awareness, but also from its formal parody of the 'conventional stance of seriousness and authority' adopted by documentary film (2014: 32, emphasis in original). Indeed, as the film's characters' 'cluelessness about how others see them' makes us laugh (with comic relief?) at the gulf between 'how we view them and how they view themselves' (Middleton 2014: 32), *Waiting for Guffman*'s playful use of direct-address interviews to frame these characters' naïve meditations makes the convention of the documentary interview itself into a joke – one that, in Middleton's view, 'subtly' erodes documentary's authoritative power (2014: 23).[10] In my own view, *Waiting for Guffman*'s deployment of puppet-like characters and mockery of documentary film's formal strategies of 'giving voice' to its subjects' authentic experience riotously skewers the ventriloquism of traditional documentary. At the same time, I argue, Guest's film emphasizes the truth-telling power of the documentary subject's *material voice* – beyond what he may or may not say to the camera. Pitch, timbre, accent, intonation and the myriad other audible properties of vocalization that constitute the (always subjectively heard) 'sound of a voice' regularly combine, in *Waiting for Guffman*, to compromise the words being spoken, endowing these words with contradictory inflections. The film's characters, in other words, speak at odds with their own voices: a Southern-twanged monotone laughably undermines a character's declaration of career-mindedness. A high-pitched, soft, sibilant, 'gay-sounding' voice makes a character's claims to heterosexuality that much funnier.

While language continues, as it always does, to signify on multiple registers in Guest's film, the carefully sounded voices of *Waiting for Guffman*'s characters play a particularly potent role in fixing these characters to identities they cannot act their way out of – no matter how hard they might try. Perhaps the film's most suggestive voice is contrived by Guest himself, who acts the part of *Waiting for Guffman*'s central character: Corky St Clair. The self-described 'actor and director and choreographer' is not, however, a recognizable iteration of Guest's own persona. Rather, Corky is an over-the-top version of the trope of the closeted homosexual: a total Broadway queen, in fact, whose flaming gayness is obvious to the viewer but remains hysterically opaque to the film's other characters – including, though of course only debatably, Corky himself.[11] Having resettled in the fictional town of Blaine, Missouri, after 'twenty-five years or so' of what one assumes was a failed theatrical career in New York, Corky sets out to direct an epic community theatre production about the town's storied history. Anticipating the rise of the reality format that would come to dominate TV programming in the early 2000s, not to mention paving the way for future generations of TV mockumentarians, Guest's film mockuments the making of this musical, entitled *Red, White & Blaine*, by 'chronicling' the (literally and figuratively) dramatic processes of casting, rehearsing and performing the show.

Through simulated observational footage and direct-address interviews with the play's participants and other townspeople, *Waiting for Guffman* lampoons

Corky's abiding pretensions to show business expertise while caricaturing the provincialism of Blaine's local 'actors' and citizens more generally. While rehearsing and performing, the cast members of *Red, White & Blaine* futilely attempt to transform their Midwestern and Southern accented voices (reflective of Missouri's ambivalent suspension between US geographical regions) into generic stage voices, a vocal failure that amplifies the Blainians' insularity. Additionally, in having the gay-voiced Corky and his regionally accented crew continually 'speak for themselves' before the camera – since *Waiting for Guffman* doesn't feature a voice-over or include an interview format beyond that of direct-address – Guest's film pokes fun at traditional documentary's alleged commitment to the sanctity of verbal testimony.[12] What Guest's characters say, in their affected voices, is usually baseless, ridiculous or both. The camera lingers on them as they lapse into awkward silences, driving home the sense that their monologues (and sometimes dialogues) are not only performances for the camera, but poorly scripted ones at that. Occasionally the camera's eye roves, cutting to 'archival' stills or b-roll footage that heightens the absurdity of the words being spoken. In Corky's case, however, nothing undercuts his speech so well as the sound of his voice, which itself screams 'gay' – even when he says nothing of the sort.

Waiting for Guffman gives us the sound of Corky's voice a few beats before synchronizing it to his sartorially eccentric and also gesturally gay image. After the film has established its small-town setting through several interviews with local residents, Corky's voice filters in over an establishing shot of a non-descript, yet clearly regional, building. 'I had been living in New York … and working there as an actor and director and choreographer for twenty-five years or so … and I *really* felt I needed a *change*,' the acousmatic voice muses, as the image-track cuts from this less-than-urbane exterior to a handheld pan across an amusing series of fake musical theatre posters. The voice is nasal, with a sing-song inflection, and its pronunciation of 'years or so' hisses on the s: 'twenty-five yearsssorssso'. In other words, the voice possesses a long 'sibilant duration': a so-called gay lisp (Levon 2006: 56). Cutting to a medium close-up of Corky upon the beginning of his next sentence, the image-track reveals the voice's speaker, whose ash blonde bowl cut, polka-dotted silk pyjama top, and flamboyant facial and hand gestures confirm our – or at the very least my – already vocally registered impression of his homosexuality (a visual confirmation not only instigated by the appearance of the tell-tale posters, but also prefigured by Corky's invocation of his migration from the consummate gay epicentre of New York to the small town of Blaine).

'I *imag*ined, in my *fan*tasy I suppose, that when I *came* here *I* would have a comp*letely* different *life*', Corky continues, his hand reaching up to twirl around the side of his head upon his sibilant pronunciation of '*fan*tasssyIsssupposse' – and flashing a wedding ring as he delicately proceeds to finger his own ear. This being the mid-1990s, and thus the solidly pre-gay marriage era, the viewer is at this precise moment made privy to one of *Waiting for Guffman*'s most pivotal jokes: Corky is not openly gay, and either is a gay-seeming married heterosexual or is unconvincingly posing as one. Either way, the effect of this disjuncture is what Middleton calls 'awkward', a word whose association with proto- or unself-conscious queerness is well documented (no pun intended).[13]

Since the voice, as musicologist Nina Eidsheim has argued, is never heard in isolation,[14] I do not intend for my analysis to suggest that the sound of Corky's voice somehow supersedes his visual appearance as an 'authentic' indicator of his homosexuality. What I am getting at, rather, is the way in which Guest's mockumentary uses Corky's direct-address to the camera – that old documentary chestnut of 'giving voice' to one's subject by allowing him to 'speak for himself' – to delegitimize the latter's speech by emphasizing his voice's gay materiality. As one subject in *Do I Sound Gay?*, David Thorpe's recent documentary on 'gay voice', puts it, Corky's 'voice is the tell' that corroborates his identity over and above whatever verbal self-description he might furnish (qtd. in Thorpe 2014). For instance, in an 'interview' that takes place after he casts *Red, White & Blaine*, Corky, clad in a Judy Tenuta 'Love Goddess' t-shirt, describes to his invisible and inaudible interviewer how his chance encounter with the wife of the local dentist prompted the latter's successful audition. In the process, his gay voice becomes even more exaggerated, its upspeak more pronounced, its hisses more drawn out: 'I was *shopping* for my wife *Bonnie* – I buy *most* of her *clothes* [*mossst*ofher*clothesss*]', he relates, quickly underscoring the random nature of the meeting's occurrence in a women's clothing store before going on to recount, 'we started talking … about panty*hose* [panty*hossse*] – she was saying … w-whatever, that's not the point of the story, but what the point is was that through this *accidental* meeting …'. As Corky references 'my wife *Bonnie*' and 'her *clothesss*' in a voice whose sound gives the lie to the language it speaks – there is no 'wife Bonnie' and the 'clothes' are not 'her[s]', but his – he loses track of his own anecdote and nearly digresses into speaking the unspeakable: 'we started talking … about panty*hose* …'. Blissfully drifting from the camera's gaze at each involuntary mention of feminine fashion, Corky's eyes collude with his 'gay lisp' to *tell on* his speech, which itself becomes more tell-tale as the tale wears on. He manages to rein it back in long enough to make his intended 'point' that actors can be discovered in the most unlikely of places: 'You *find* people, you *find* them, something – is it karma? Maybe. But we found him; that's the important thing'. But the 'interview', as such interviews often do, ends on a punch line. As the camera inches closer to Corky's face, he delightedly adds, '*And* I got Bonnie a *wonder*ful *pant*suit'.

This sequence is especially interesting because it *outs* Corky not despite, but rather *through*, his own explicit references to the previously unnamed 'Bonnie'. Corky's outing isn't merely rhetorical, in a 'the lady doth protest too much, methinks' kind of way. Nor is it simply accomplished by his suggestive references to pantyhose and pantsuits. His outing is largely accomplished by his voice, which, in its inability to stifle his thrill at the eventual discovery not of a new actor for his play but of a '*wonder*ful *pant*suit' for 'Bonnie', betrays the fact that the pantsuit is actually for him. Furthermore, the camera's gradual approach to the source of this indirect yet piercingly clear declaration of gayness 'subtly' underscores – to return to Middleton's earlier point about Guest's micro-subversion of traditional documentary film form and its attendant truth claims – the scene's constructedness, pointing up Guest's (i.e. the actor and filmmaker's) choreography of the punch line, whose very punchy delivery is what clinches the joke. Thus, when Corky later, and with characteristic bitchiness, tells his passive aggressive musical director to 'just talk like a normal person' instead of speaking in an angry whisper, the joke is as much on the musical director as it is on Corky, since neither is supposed to sound 'normal' – whatever that would mean. And when we find out, in a direct-address interview at the film's close, that Corky has returned to New York to audition for the role of 'the somewhat stern taskmaster … who teaches Liza how to speak "correctly"' in a *My Fair Lady* revival, the coincidence is too perfect. For in addition to the gay rumours ever swirling about the character of Henry Higgins,[15] Corky's demonstrated attempt to perfect Higgins's English accent is a marble-mouthed failure. The material of Corky's voice is not supple enough to be sculpted otherwise, and his inability to hear his own voice's stubborn lack of malleability again secures the joke.

As its final scene attests, *Waiting for Guffman*'s 'awkward aesthetic' arises as much from the chasm between how we *hear* the film's characters and how they *hear* themselves as it emerges from the perceptual gulf between 'how we view them and how they view themselves' (Middleton 2014: 32). This hearing, I have argued, has as much to do with sensing the materiality of a character's voice as it does with interpreting the significance or referential content of their linguistic speech. Indeed, the character of Corky St Clair uproariously exemplifies how voice can, and often does, diverge from speech to 'speak the truth' – or, in Corky's case, to speak 'the love that dare not speak its name'. To be sure, Corky's speech alone contains enough latent – or not so latent – queer content to clue us into his open secret (a stray invocation of 'chaps' here, a 'pantsuit' mention there), but his stereotypically lilting, lispy 'gay voice' simultaneously amplifies these references and relegates 'Bonnie' to the realm of fantasy beard-dom. That Corky cannot, in the end, ventriloquize Higgins, the consummately urbane British homosexual, is the ultimate irony. As he performs his concluding direct-address 'interview' before the 'documentary' camera, we're made painfully aware that for all Corky's queerly voiced claims to worldliness, this theatre queen belongs in Blaine.

'We were speechless': *Modern Family* ('Yard Sale', 2012)

A 2012 episode of the mockumentary sitcom *Modern Family* directly connects the trope of 'gay voice' so central to Guest's film with the trope of the ventriloquist's dummy to hint at, but ultimately disavow, the documentary subject's ventriloquism. Like many television mockumentaries produced in the present heyday of reality TV, *Modern Family* features documentary-style direct-address 'interviews' with its central characters, who in the mock observational footage that comprises the rest of each half-hour-long episode, more subtly register their continual awareness of being filmed by knowingly or guiltily locking eyes with the camera's anonymous gaze.[16] The clan whose 'modern' life the show chronicles is a Los Angeles-based blended family helmed by an older white American patriarch with a young, heavily accented Colombian second wife and stepson. His adult biological daughter and son each have families of their own: the former directs an updated straight white 1950s-style ensemble while the latter, who is openly gay, has a flamboyant white partner and an adopted Vietnamese daughter. *Modern Family* parodies the quotidian miscommunications that arise as this sprawling crew, possessed of exaggeratedly strong and silly personalities, negotiates its own internal differences. Like many non-documentary-style family sitcoms, the show errs on the side of sentimentality. In *Modern Family*'s would-be transparent direct-address segments, its characters are just as likely to verbalize stirring 'morals of the story' as they are to voice their own laughable self-opacity. The episode 'Yard Sale' (Episode 6, Season 4), first aired in 2012, is thus notable for explicitly thematizing both 'gay voice' and ventriloquism – which at first listen seems like a meta-commentary on the show's use of documentary conventions to extract its fictional subjects' genuine voices – while emphasizing the truth-value of direct-address.

The episode's guiding premise is that the family is throwing a yard sale chez Jay (the patriarch) and Gloria (his wife) to raise money for a charity fundraiser organized by Manny and Luke's (Jay's stepson and grandson's) class at school. One plotline involves the two boys' comical discovery of a ventriloquist's dummy as they search for items to sell at the yard sale; the puppet, it turns out, belongs to Gloria, who has closeted it away because it recalls a shameful event from her Colombian past. A second plotline involves a more metaphorical closeting: Claire (Jay's daughter and Luke's mother) and her nerdy daughter Alex (Jay's granddaughter and Luke's sister) clash when Claire becomes convinced that Alex's new boyfriend is leading her on. Indeed, the sound of the boyfriend's voice has Claire 'ninety percent sure' her daughter's new beau is 'a hundred percent gay',[17] so when Alex insists on inviting him to the yard sale, Claire gets her gay brother (and Alex's uncle) Mitchell to step in and convince his niece of her boyfriend's homosexuality – a botched task, to say the least. Both plotlines feature the 'outing' of a character who has failed, or simply refused, to say something out loud: Gloria has not confessed her secret ventriloquism; Michael (the gay boyfriend) has not confessed his queerness. But while the episode gives the thickly accented Gloria the didactic power to explain the dummy's concealment to the rest of her family – and subsequently perform a redemptive puppet show – it does not extend the privilege of self-representation to Michael, foregrounding instead the family's external assessment of his sexuality on the basis of his voice's sonic materiality. This has as much to do with the limitation of the episode's (and the series') direct-address segments to members of the family as it does with the show's apparent conceit that only the family is being documented. For on *Modern Family*, character awareness of the camera filming the family's life seems limited to the family's own members; incidental figures like Michael remain more or less ignorant of its presence. The show thus demonstrates the extent to which 'mockumentary has become naturalised within the television lexicon' (Hight 2010: 197), exemplifying contemporary television's increasing use of documentary and reality forms as 'simply [some] of many aesthetic options available in contemporary media culture' (Middleton 2014: 19) – rather than as means to critique documentary media's way/s of representing reality.

To wit, the 'gay voice' plotline in 'Yard Sale' begins with observational footage of Claire's uneasiness at Alex's mention of her boyfriend, after which are intercut two brief direct-address segments that quickly contextualize the interaction: Claire tells the camera she thinks Michael is gay, and Alex tells the camera she doesn't (because Michael 'made out' with her). Later on, in observational footage of the yard sale, Claire apprises Mitchell and his partner Cam of her discomfort with Alex's relationship, at which point Alex and Michael approach the trio to greet them. The older gay couple, whose joint queeniness is, like Gloria's accentedness, one of the series' comic cornerstones, makes the acquaintance of the allegedly nascent young gay, the voice of each rising in pitch as their 'Hi' and 'Hello' are elongated into multisyllabic utterances. Smiling and responding in kind, Michael draws out his own (relatively less) high-pitched 'Hi' ('Hieeeeee'), following the salutation with a bashful intake of breath before running off with Alex. The boy's brief, singular utterance serves as instant confirmation of his homosexuality. 'Yup, he's gay,' quips Mitchell, giving Claire's earlier assumption a homosexual stamp of certitude. Claire then looks directly at the camera as if to imply, 'I told you so'. Shortly thereafter, Claire, Mitchell and Cam delicately attempt to inform Alex of Michael's gayness, to which Alex responds, 'It's not the olden days anymore, you guys; I can just ask him'. And she does. In the observational footage of their conversation, Michael, clearly ruffled, insists, 'I'm straight' – with a classic gay lisp. 'Are you a boy?' he asks Alex. 'No', she replies. 'Right, so [ssso] then I'm straight [ssstraight].' Unaware of the camera's presence, Michael angrily asks to be left alone. Alex glances up at the camera as she leaves him be, indicating surprise at the violence of his denial. The two are later shown to have made up, with Claire, Mitchell and Cam observing that the relationship will not end well – but that they must let it run its course.

What's so striking about the manner in which this plotline unfolds is the mockumentary's formal refusal to let Michael have a meaningful linguistic voice as it confirms his material voice's obvious conveyance of a singular identity: gay. While Claire, Mitchell, Cam and Alex's voices form words whose meanings aren't belied by their sound, Michael's words are untrustworthy – his protestations unsound, as it were. The joke is on him, and nearly everyone but him is in on the joke. And the joke, which also functions as a moral, seems to be that one can't be outed before one's time: it's a patronizing life lesson about the years it can take to come to terms with one's sexuality (i.e. to 'find one's voice'), rather than a parody of the 'gay voice's' often contradictory linguistic and material claims to essence.

In contrast, 'Yard Sale' eventually gives Gloria (played by Colombian-American actress Sofía Vergara) the ability to respond both expositionally and playfully to her ventriloquist dummy's un-closeting. When Manny and

Luke's unearthing of the puppet is brought to her attention (in an observational sequence involving multiple family members), she immediately launches into an explanation of its provenance despite her visible shame. As the image-track cuts away to an 'archival' still photograph of a row of bikini-clad young women – a documentary technique rarely used on *Modern Family* – Gloria recalls learning ventriloquism as her 'talent' for a Colombian beauty pageant. Her accented voice, whose racialization the show regularly exploits for laughs, dramatically rises and falls as she recounts that when it came time to perform with her dummy, Uncle Grumpy, 'I froze, and we were speechless'. Gloria insists that she will never practice ventriloquism again due to this previous 'humiliati[on]', but Luke cannily manipulates her into performing before the family by implying that her devoted son Manny will lose respect for her if she succumbs to cowardice. In a brief direct-address segment, Luke makes clear that *he* is the real puppet master, declaring, 'I know how to push buttons to get what I want. And what I want is to see a puppet show'. Luke's apparent ease in goading his Colombian step-grandmother into presumed public humiliation through his invocation of filial piety dovetails with the show's presentation of Gloria as hilariously unaware of the idiosyncrasies of her own English-speaking voice, racializing her into the butt of the mostly white family's joke. As we learn that Gloria will indeed ventriloquize for the entertainment of all, she explains her change of heart via direct-address. 'Luke's wise words got to me,' she earnestly concedes. 'I needed my son to respect me again. More than that, I needed to respect myself.' The assumption, of course, is that the opposite will occur, with Gloria's puppetry culminating in certain shame and disgrace.

When Gloria finally performs her impromptu ventriloquist's act for her delighted family, however, the 'respect' she so craves materializes (see Figure 8.1). Rather than humiliating herself, as Luke and everyone else felt her sure to do, Gloria, aided by the crass Uncle Grumpy, turns the tables, even making jokes at her mostly white family's expense. Contorted into a lower-pitched, gravelly version of its typically accented self, her voice, channelled through the mediating figure of the sour-faced and bushy-haired 'male' dummy, shouts, 'Ai, look at all those white faces! I see we made it over the wall!' Gloria's shift from 'speechless' beauty pageant contestant to powerful ventriloquist is affirmed by the rest of the family, whose members laugh at her jokes despite themselves. Although they expected to be laughing at her, and not at her comedic prowess, her shrewd instrumentalization of her own accented voice, her crafty repurposing of her voice's supposedly laughable materiality, gets Gloria the last laugh.

Figure 8.1 Gloria (Sofia Vergara) entertains the family with her un-closeted ventriloquist's dummy, Uncle Grumpy, in 'Yard Sale' (dir. Gail Mancuso). Season 4, Episode 6, *Modern Family* (ABC, created by Christopher Lloyd and Steven Levitan) 31 October 2012. Screenshot CityTV, Toronto, Canada. Rogers Media.

Thus, while Gloria, aided by the documentary convention of (always deceptively) direct-address, is permitted a redemptive rediscovery of the literal and metaphorical voice she thought she'd closeted away forever – as well as, perhaps, the discovery of a newly ventriloquial iteration of her voice? – 'Yard Sale' implies that Michael must grow to accept the stable identity his voice's ostensibly 'gay' materiality has prescribed for him. Of course Gloria's palpable accent will, throughout the rest of the series, continue to serve the purpose of fixing her to a seemingly authentic racialized subjectivity, but the episode under discussion packs a brief break from that dominant tendency when it allows Gloria's ventriloquial 'grumpy' voice to speak the unspeakable whiteness of the not-so-modern family. As a queer outsider to the latter, Michael is more object than subject of the mock 'documentary' that acts as *Modern Family*'s organizing principle. Like that of Corky St Clair, Michael's voice sounds at odds with his speech, but he has no play, no performance, no ventriloquist's dummy through which to explore his voice's obstinate or plastic materiality – just the unheard assertions of a few knowing adults who say he'll come to terms with the voice he's been given when he's ready.

Coda

The case studies in this chapter perform different readings of the 'voice of mockumentary' – a voice that is of course not singular but plural, encompassing as it does a range of material and metaphorical voices. Rather than drawing a hard and fast distinction between the (compromised) voice of the documentary subject and the dominant 'voice of the text' – a distinction that metaphorizes the subject into a dummy and the text into a ventriloquist (to say nothing of its erection of a false binary to begin with) – I have focused on the relation, within each mockumentary text, between vocality and documentary convention. In my attempt to describe how these texts, which set out to 'mock' documentary, redefine the role of the voice as a central authenticating device therein, I have argued that while mockumentary often self-consciously positions the voice as ventriloquial, it also nevertheless relies heavily on the voice's material and linguistic authority. Double-tongued, the voice of mockumentary does what it has to do to secure the laughs on which it thrives; thus, its usage of documentary conventions of voicing is never simply 'complicit' or 'subversive'. In the end, in *Waiting for Guffman*, the joke of Corky's gay voice is not merely made *on* this character but *by* him, in an important sense because the character is played by the filmmaker himself. And though Guest's rendition of the voice of Corky St Clair suggests that the truth of the latter's sexuality inheres in the sonic materiality of his voice (his sing-song inflection, his wide pitch range, his prominent lisp, his voice's many other audible properties), we know this rendition to be exploiting a trope, not making a truth-claim. *Modern Family*'s 'Yard Sale' episode, for its part, hews more sincerely towards the popular notion that proper voices do exist and can be 'found', dramatizing the dichotomy between self-aware, 'gay-voiced' elders and an awkward youngster whose audibly 'gay voice' tells a truth he isn't ready to admit. However, through its incorporation of the trope of the talking puppet – that most uncanny of narrative devices – 'Yard Sale' implies that all of our voices are more ventriloquial than we might like to admit.

Notes

1 My title, of course, refers to that of Bill Nichols's well-known essay, 'The Voice of Documentary' (1983).

2 I explore the notion of the documentary's voice's ventriloquism at length in a recent essay in *Camera Obscura* (Kessler 2016). For recent theorizations of the ventriloquism of the voice itself, see Connor (2000), Žižek (2001) and Dolar (2006).

3 See Gaines and Renov (1999).

4 My preference for the term 'mockumentary' over 'fake documentary' (see Juhasz and Lerner 2006: 1–35) or 'mock-documentary' (see Roscoe and Hight 2001: 1–5) is aligned with Alisa Lebow's, as expressed in her essay 'Faking What? Making a Mockery of Documentary' (2006). She writes, 'I believe the term "mockumentary" more effectively works to signal a skepticism toward documentary realism, rather than to reauthorize documentary's "truth" against the fake doc's "false." Like "fake doc," the term "mockumentary" incorporates and implicates *documentary* into its referent while still implying some distinction from it' (Lebow 2006: 224, emphasis in original).

5 On documentary and 'reality', see, among others, Nichols (1992) and Gaines and Renov (1999: 1–17).

6 Or should I say, 'documentary *studies*' tendency to collapse the capacious category of "voice" into linguistic testimony'? It is not this chapter's intention to create a false dichotomy between documentary and mockumentary forms, such that the former always already conflates 'voice' with language and speech while the latter always already engages the voice's materiality – with both forms still often working towards similar authenticating ends. Although dominant iterations of documentary form do indeed continue to construe speaking as voicing (and vice versa), it is documentary *studies*, as Pooja Rangan brilliantly argues in her monograph *Immediations*, that remains almost singularly 'invest[ed]… in the speaking voice as a measure of humanity,' to the exclusion of other ways of hearing documentary's proliferating voices (2017: 134).

7 See Schlichter and Eidsheim (2014), for a compelling description of what voice studies, as a transdisciplinary field of inquiry, might (and now, I think, does) look – or rather sound – like. In addition to the scholars and works cited by Schlichter and Eidsheim, both the present volume and a recent special issue of *Discourse* on voice in documentary (Rangan and Yue 2018) make crucial contributions to the field.

8 Middleton's definition of 'awkwardness' is indebted to Kotsko (2010).

9 Within linguistics, the literature on 'gay voice' is extensive, with much of it focusing on listeners' abilities, or lack thereof, to correctly identify a speaker's sexual orientation on the basis of the sound of their voice (Smyth et al. 2003). As Levon points out, 'gay voice' is often popularly associated with 'pitch range and sibilant duration', both elements of prosody or patterns of stress and intonation (2006: 56). In other words, in addition to being high-pitched, the stereotypically construed gay male voice has a wider pitch range than the stereotypically construed straight male voice. The former is also popularly understood to hiss more on its s-sounds (i.e. to have a longer sibilant duration), thus creating a so-called gay lisp (Levon 2006: 56).

10 I am applying Middleton's formulation of 'awkward aesthetics' to *Waiting for Guffman*. He does not analyse this particular film in his book.

11 The queer literature on the closet/closeting/closeted-ness is extensive. For paradigmatic examples, see Sedgwick (1990) and Halperin (2002). For a critique of the metaphor's use in queer theory, see Ross (2005).

12 It should be noted that all of Guest's films mock documentary's stock in verbal testimony – not only *Waiting for Guffman*.

13 In the realm of queer theory, see Sedgwick's essay, 'Shame, Theatricality, and Queer Performativity: Henry James's *The Art of the Novel*', in which, among other things, she discusses shame as affect vis-à-vis James's *The Awkward Age* (2003: 38–47).

14 See Eidsheim (2012: 13). Eidsheim writes, 'voice is never *heard* in a state prior to the impact of cultural, social, and other outside forces'.

15 See Ray (2014: 300–302).

16 *Modern Family* is a single-camera, laugh-track-less series. The fictive nature of its 'observational footage' is of course given away by the formal presence of shot-reverse-shot and other editing techniques that create a space that appears apparatus-free.

17 Claire delivers this punchy statement in a brief direct-address session with the camera.

References

Bayer, G. (2006), 'Artifice and Artificiality in Mockumentaries', in G. D. Rhodes and J. P. Springer (eds), *Docufictions: Essays on the Intersection of Documentary and Fictional Filmmaking*, 164–178, Jefferson, NC: McFarland and Company, Inc.

Connor, S. (2000), *Dumbstruck: A Cultural History of Ventriloquism*, Oxford: Oxford University Press.

Dolar, M. (2006), *A Voice and Nothing More*, Cambridge, MA: MIT Press.

Eidsheim, N. (2012), 'Voice as Action: Toward a Model for Analyzing the Dynamic Construction of Racialized Voice', *Current Musicology*, 93: 9–32.

Gaines, J. M. and M. Renov, eds (1999), *Collecting Visible Evidence*, Minneapolis, MN: University of Minnesota Press.

Halperin, D. M. (2002), 'Homosexuality's Closet', *Michigan Quarterly Review*, 41 (1): 21–54.

Hight, C. (2010), *Television Mockumentary: Reflexivity, Satire, and a Call to Play*, Manchester: Manchester University Press.

Juhasz, A. and J. Lerner, eds (2006), *F Is for Phony: Fake Documentary and Truth's Undoing*, Minneapolis, MN: University of Minnesota Press.

Kessler, S. (2016), 'Puppet Love: Documenting Ventriloquism in Nina Conti's', *Her Master's Voice, Camera Obscura: Feminism, Culture, and Media Studies*, 31 (2): 61–91.

Kotsko, A. (2010), *Awkwardness: An Essay*, Washington, DC: Zero Books.

Lebow, A. (2006), 'Faking What? Making a Mockery of Documentary', in A. Juhasz and J. Lerner (eds), *F Is for Phony: Fake Documentary and Truth's Undoing*, 223–238, Minneapolis, MN: University of Minnesota Press.

Levon, E. (2006), 'Hearing 'Gay': Prosody, Interpretation, and the Affective Judgments of Men's Speech', *American Speech*, 81 (1): 56–78.

Middleton, J. (2014), *Documentary's Awkward Turn: Cringe Comedy and Media Spectatorship*, New York: Routledge.

Minh-Ha, T. T. (1992), *Framer Framed*, London: Routledge.

Nichols, B. (1983), 'The Voice of Documentary', *Film Quarterly*, 36 (3): 17–30.

Nichols, B. (1992), *Representing Reality: Issues and Concepts in Documentary*, Bloomington, IN: Indiana University Press.

Rangan, P. (2017), *Immediations: The Humanitarian Impulse in Documentary*, Durham: Duke University Press.

Rangan, P. and G. Yue, eds (2018), 'TK. Audibilities: Voice and Listening in Documentary', *Discourse: Journal for Theoretical Studies in Media and Culture*, TK.

Ray, M. (2014), '*My Fair Lady*: A Voice for Change', *American Music*, 32 (3): 292–316.

Roscoe, J. and C. Hight, eds (2001), *Faking It: Mock-Documentary and the Subversion of Factuality*, Manchester: Manchester University Press.

Ross, M. B. (2005), 'Beyond the Closet as Raceless Paradigm', in E. P. Johnson and M. G. Henderson (eds), *Black Queer Studies: A Critical Anthology*, 161–189, Durham, NC: Duke University Press.

Schlichter, A. and N. S. Eidsheim (2014), 'Introduction: Voice Matters', *Postmodern Culture*, [e-journal] 24 (3). Available through: UCI Libraries website http://lib.uci.edu (accessed 28 July 2017).

Sedgwick, E. K. (1990), *Epistemology of the Closet*, Berkeley, CA: University of California Press.

Sedgwick, E. K. (2003), *Touching Feeling: Affect, Pedagogy, Performativity*, Durham, NC: Duke University Press.

Smyth, R., G. Jacobs, and H. Rogers (2003), 'Male Voices and Perceived Sexual Orientation: An Experimental and Theoretical Approach', *Language in Society*, 32 (3): 329–350.

Do I Sound Gay? (2014), [Film] Directed by David Thorpe, USA: Impact Partners.

Waiting for Guffman (1997), [Film] Directed by C. Guest, USA: Castle Rock.

Yard Sale (2012), *Modern Family*, [TV programme]. ABC, 31 October 2012.

Žižek, S. (2001), *On Belief*, New York: Routledge.

9

Fabulous Thinking, Chimeric Voice: Kathy High's *Lily Does Derrida: A Dog's Video* (2010)

Brenda Hollweg

In memory of Pancho (2001–13)

'I am … *was* a dog', the eponymous Lily from Kathy High's *Lily Does Derrida: A Dog's Video* (2010) corrects herself. She does this in the same way any human speaker, if self-reflexive enough, would be able to do. But Lily is a canine, 'a street dog' that 'lived with a human and two cats' and came to High 'by chance', as we learn from her own account: 'She and I had some fun over the years. She thinks she understands me. Maybe she does, better than most. I think I understand her too. We talk a lot about our day's events. She lets me complain which I do a lot lately, and sometimes we just enjoy sitting together. I know you all think this is anthropomorphizing me'.

The impression that Lily possesses the power of human speech is obviously not a result of her own reasoning and vocalizing abilities, but rather the effect of a complex cinematic fabrication. Live animals, apart from parrots and a few other birds, do not speak or sing in human languages. If they appear to do so on film, this is usually made possible through work in post-production, when filmmakers separate audio from video tracks, juxtaposing or recombining them to manufacture mixed beasts, composite figures or chimeras with the body of a live action animal and the voice-over of a human. This voice-over is often carefully scripted and performed by professional voice actors, producing a mixture of artifice and realism that cuts across species barriers.[1]

I am interested in the possibilities such voice fabrication offers High as cine-essayist in developing a more fluid and democratic notion of trans-species communication and collaboration. In the classical anthropocentric model, a lack of language 'has long been seen as not just a sign of difference, but as the ultimate sign of animal inferiority to humans' (DeMello 2013: 5). By comparison, High's use of a beast that speaks with a chimeric voice allows for a shift in perspective from a human-centred model to one in which a perceived equality between human and non-human animals makes it possible for the latter to acquire what Joan Gordon calls a 'contoured subject status' or 'personhood' (2010: 459). In her study on speaking animals in science-fiction literature, Gordon reworks Haraway's cyborg and Jacques Derrida's *animot* to coin the term *amborg*, a human–animal interface, in which polarities and hierarchies, including those between nature and culture, human and animal, fiction and social reality are at issue. The chimeric voice that High creates for her mythical beast in *Lily Does Derrida* is the non-locatable voice of the hybrid/interface amborg. This voice does not originate from a particular human body, although human voice-over/off-screen narration is a part of it; nor does it emanate from a non-human animal body, although certain prelinguistic sounds may be part of it. The chimeric voice is the effect of a more complex cinematic projection.

When High fabricated this voice in post-production, her pet dog Lily was already dead. But through cinematic projection, Lily is given a post-mortem existence as an imaginative creature who appears to have an innate power of reason, is versed in philosophical discourse and also practises other-than-human speak. As the main character of the narrative Lily talks in a language we understand and gives us an account of herself, even of her own death. She can also comment on her relations with High, her human guardian, as well as with other humans and animals (cats, in particular) or ponder the state of human–animal affairs. She both argues with and challenges Derrida's philosophical work on human–animal relations; and in doing so, she provides her listeners with conflicting viewpoints and takes up shifting speech positions. Some of these positions allow her to reflect in partly humorous, partly ironic ways, on her fabricated existence as anthropomorphized chimera. Jarring with this narration, however, are images of Lily, the live animal, in action: images that project back to us a 'consciousness' that appears to be oblivious to the intricacies of abstract human thought (see Figure 9.1). This dog's main interests seem to lie in the acquisition of food, in perceiving the world through scent rather than sight; being a sentient being, her life, like ours, is finite, and she shares profound experiences with us, such as joy, affection and pain. This existence of the projected Lily as a creature of fiction as well as of social reality, with a capacity to affect and be affected, is what accounts for the impure, polymorph and polytemporal nature of her chimeric voice.

Figure 9.1 Versed in trans-species communication: High's canine philosopher 'Lily'. *Lily Does Derrida: A Dog's Video* (Kathy High 2010). Courtesy Kathy High.

The way High deploys this voice to engage in what Haraway calls 'multispecies storytelling in the feminist mode' (2013: n.p.), which in the cine-essay always also constitutes a material-semiotic mode of heresy, is qualitatively different from a wide range of chimeric voice fabrications in film, on television or video-sharing websites such as YouTube. The gap that is repeatedly opened up in *Lily Does Derrida* between voice/voice-over/off-screen narration and image invokes a form of self-reflexive irony that can also point us in a meta-referential way to the conditions under which this creature comes into existence.

Chimeric voices across a diversified media landscape

Whenever live action animals appear to reason or crack a joke in a human language, they speak with a chimeric voice.[2] This is even the case in family dramas or adventure films for young audiences in which animal characters are often heavily anthropomorphized. Like human actors, these characters wear clothes, play instruments, sing or engage in different games. They are also made to behave and speak like humans, due to the work of animal trainers, special dubbing and lip-syncing techniques or computer-generated imagery. The orphaned piglet in *Babe* (1995), the wrestling Jack Russell Terrier in *Russell Madness* (2015) or the soulful Basset Hound from *Pup Star* (2016), who teaches Yorkshire Terriers to sing the blues, all constitute speculative voice chimeras whose actual existence *as* live action animals is rendered invisible through such devices. This offers audiences, particularly if they are relatively young, opportunities to fantasize, sympathize or dis/identify with the central animal characters. 'Children often don't recognize', as Kathy Rudy points out, 'that the rest of us think of this phenomenon as an anthropomorphized fantasy. Indeed, it is a right [*sic*] of passage into adulthood to accept the "reality" that only humans have language' (2013: 149).

Only occasionally – as in Barry Sonnenfeld's *Nine Lives* (2016) where a mixture of live animal action and CGI-imagery is deployed to create the fantastic TomCat – the dialogue works more explicitly with semantic gaps, humour and irony to raise an awareness of the paradoxical conditions under which an animal can speak in a human language. These rhetorical structures in the narration itself account for, in Haraway's words, 'the *pleasure* in the confusion of boundaries and for *responsibility* in their construction' (1985: 66).

As film phenomenologist Vivian Sobchack (2004) argues in her study on embodiment and moving image culture, however, live animals – even those appearing in fiction film – always bring in a sense of the real and actuality to an otherwise fictional scenario.[3] In many of these productions this element is often used to enhance the thrill of a particular scene or to serve commercial interests, even as it is exploited at the expense of the live animal. In Duwayne Dunham's *Homeward Bound: The Incredible Story* (1993), a family adventure in which the different human voice-overs are loosely projected onto the on-screen action of the animal actors, this tension between artifice and realism becomes more visible. The dialogue requires Sassy (Sally Field), a Himalayan cat and one of the central animal characters, to cry for help in human language while we watch the feline actor (or what appears to be a live animal) being carried away by strong river currents that eventually push her down a waterfall. The cinematographically produced realism of this life-threatening situation renders the existence of Sassy *as* live animal/cat performer suddenly visible.

In fiction films in which animal characters function as narrators or central focalizers of the story – such as the series of re-incarnated Bailey-dogs in Lasse Hallström's *A Dog's Purpose* (2017) or the street-wise underdog in Vladimir Bortko's critical allegory on the Soviet regime, *Heart of a Dog* (1988) – other-than-human speak becomes more visible and audible as a constitutive element of the chimeric voice projection. This is due to the use of narrative techniques of interiority in these films, including interior monologue and stream-of-consciousness. Such techniques make the human voice fade into the background to the extent that the specific ways the live action animals communicate in species- or even breed-specific ways (for instance, through the way they sit, play, walk, howl or whimper) can be registered.

The initial quotation from *Lily Does Derrida*, which ends with Lily's tongue-in-cheek remark, 'I know you all think this is anthropomorphizing me,' displays a form of self-awareness and self-reflexive irony that is in a sense typical of cynomorphic tales. One example from early European literature is *El Coloquio de Los Perros* (Dialogue of the Dogs), a novella written by Miguel de Cervantes as part of a series of short novels written between 1590 and 1612. It includes a scene in which two canines reflect upon their capacity to speak human language and to invent creatively. To provide these dogs with a distinct personality Cervantes makes use of the narrative strategies and character formation of the picaresque novel, which typically adopts a realist style deflected by elements of comedy and satire. In their perceptiveness, shrewdness and capability to serve multiple masters, Cervantes's canines call up the *pícaro*, a figure of low standing, who rarely deigns to hold a job but gets by with brash bearing and wit.

Transgressive animal characters that are gifted with the power of human speech are also known from American film and television of the 1950s and 1960s. Arthur Lubin's comedy drama *Francis* (1950) in which an old army mule befriends a young, naïve GI, or the American television production *Mister Ed* (1961–68), which Lubin also co-directed, were highly popular at the time. The successful launch of *Francis* inspired Universal Studios to produce a whole series of further Francis-films between 1951 and 1956. Here, as well as in the later sitcom classic *Mister Ed*, the hoarse and textured voices of actors Theodore Childress Wills (Francis) and Allan Lane (Mister Ed) carry the picaresque, at times blasphemous bite that is the trademark of these loquacious equines. Lily's voice chimera is less brash, as she tackles in a more reflective and widely read manner major human voices in Western philosophy. But her role as subaltern bitch, which takes on binary structures and hierarchies, resonates with these quick-witted characters from early American film and television, which likewise aim to engender a more critical perspective on the human condition as well as the use of different language systems.

Over the last decade an increasing number of live animals that also speak in the human language have begun to roam the off- and online reality formats of comedy sketch shows, advertisements, personal blogs or vlogging sites. In BBC One's *Walk on the Wild Side* (2009–) scriptwriters, actors, voice-artists, comedians or radio and television hosts make use of special performance techniques to dub wildlife filmed in their natural habitats. Most of these techniques are known from stand-up comedy, voice impression or animation. In the late 1980s, the British animator Nick Park (Aardman Animations) came to much acclaim when he synced the lip and facial movements of his stop-motion clay-modelled zoo animals with voice-over commentary recorded at housing estates and retirement homes in the UK. The aim was to make these clay figures appear as if they were being interviewed about their living conditions. In *Creature Comforts* (1989), jaguars, turtles and gorillas talk with the regional and national dialects, accents and inflections of human speakers. In a similar way the dubbing in *Walk on the Wild Side* provides duck-billed platypuses, mudskippers or barn owls with explicitly human-centred gender, race and class stereotypes and a prime-time television audience with its weekly dose of cross-species entertainment.[4]

On video-sharing websites like YouTube a whole universe of user-produced material exists in which live animals are heard to communicate and seen to co-perform with their human guardians.[5] The imagery deployed in many of these works has usually been captured spontaneously and in an eclectic way, and audiences are provided with social rather than professional actors – typical of the work of the amateur or self-producer. To some extent these short films form videographic 'slice-of-life' stories: documents of the ordinary and the mundane, gleaned from predominantly white and middle-class settings at home, but also at zoological gardens around the world. In various cases subtitles are added that creatively translate the many different sounds the on-screen animals make into human language.

Again and again the apparent similarities and explicit differences between species and their communicative systems are highlighted. Many of these scenarios bring to light what Karla Armbruster describes as some of the most narcissistic and anthropocentric tendencies within Western culture. In 'What Do We Want from Talking Animals?' (2013) Armbruster looks at material from a CBS news feature that appeared online on 4 April 2013. In this feature, a Pug, a Siberian Husky, and other loquacious dog breeds express themselves in a language system that is very different from our own, but for their owners these sound expressions become phrases such as 'I love you'. In her essay Armbruster also alerts us to the crudely sentimental and/or self-adulating ways in which these cross-species encounters are often (mis)represented by the edutainment industry.[6]

Lily Does Derrida sits somehow in between these diverse cinematic genres, from which it borrows in an eclectic way. The malleable form of the cine-essay makes use of narrative techniques of interior monologue or stream-of-consciousness known from literature and fiction film as well as dubbing strategies associated with animation to provide Lily with a distinct vocal personality. Lily also speaks with a voice that is as self-reflexive and meta-referential as that of Cervantes's dogs, and as irreverent and tongue-in-cheek, albeit perhaps less vulgar, as that of the loquacious equines from early American film and television. The humour and self-directed irony that she deploys to make us aware of her condition as anthropomorphized voice chimera is made to expose and challenge a larger trend in popular culture and media to appropriate and thereby objectify the existence of live animals as entertainment machines.

Cynomorphic heresy: Kathy High's cine-essayistic voice chimera

Working in documentary and experimental film for over twenty years, High has produced full-length video essays and mixed media installations that explore the ethical and political ramifications of democratic forms of cross-species communication and collaboration. In *Animal Attraction* (2001) and *Lily's Nightmare* (2003), a shorter piece about her pet's anxiety about lightning, Lily and several farm animals are assigned contoured subject roles and ascribed agency comparable to those of human collaborators. While they are mentioned in the film credits as 'creators', High's name appears as 'collaborator' or 'facilitator'. Although the filmmaker seems to acknowledge an ontological difference between human and non-human animals, the rhetorical structure of her video essays seeks to change the hierarchical and binary logic that is traditionally inscribed in this relationship. By means of creating a chimeric voice for Lily,

High does not aim to give speech to 'the animal' but to practise a more fluid form of becoming-with; a 'solidarity-building projection of other creatures' ways of being-in-the-world' that also enables 'a reassessment, in turn, of forms of human being' (Herman 2016: 1).

When *Lily Does Derrida* opens with a close-up shot at low angle, we see the dog with eyes first closed and then half-opened lying on the floor, dozing and snoring. The scene is followed by a sequence of medium-long shots that have the visual quality of flashbacks or thought fragments seemingly passing through the dog's mind. This speculative scenario calls up a scene from the beginning of Charles Siebert's novel *Angus* (2000) in which the eponymous Angus, a Jack Russell terrier, reminisces about his life as he lies dying. Through his interior monologue, which 'switches from registering present sensations to recollecting episodes from the past' (Kuzniar 2006: 59), the phenomenology of the dog is rendered visible. Imagery of Lily hobbling along a wooded path or trudging through muddy puddles is deployed to similar effects. The dynamic, unsteady camcorder aesthetics of High's video essay underpin the creature's joyful approach to her surroundings, but also stress her wobbly gait and limited vision as an ageing dog.

The opening sequence is followed by visual material that shows the dog resting again; but this time her eyes are open, clearly indicating that she is awake. Through a voice-over narration, written for her by High, Lily's supposed thoughts, comments and daily observations become audible. In this way a speculative voice character ('Lily') is created for the dog. 'There is a lot of pressure on me', we hear this character say in a laconic voice: 'A lot is riding on me now. And time is short.' In *Melancholia's Dog* (2006) Kuzniar argues that the canine narrators or central focalizers of cynomorphic tales often show a high degree of self-awareness; and this is also strongly suggested in *Lily Does Derrida*. From High's documentary imagery we learn that the actual pet dog is an ageing street dog, slowly but surely going blind. Her speculative character, however, has discovered the value of books: 'I've been reading a lot these days. Jacques Derrida's caught my interest. Too bad he died, but then – I'm dead too, by the way.' This 'other' Lily knows that she – like the French philosopher – is already deceased when the film was made. So, when Lily can be heard to speak in a high register, philosophize over human–animal relations or ponder the possibilities of an afterlife, viewers also become aware of the irony that they are actually listening to Lily's post-mortem voice, retrospectively fabricated for her by High.

Although the Lily character denotes a female dog, the voice she speaks with in human language is that of voice actor Josh Thorson. His male timbre and US-American accent strangely situate and transgender the animal character; but through a male voice she might be in a better position to strike a pose as philosopher, to gain authority and 'do', as the film title suggests, Derrida. This does not mean that she talks *like* Derrida in the sense of voice impersonation, but rather thinks *with* him about the world and human–animal relations, argues alongside his work, from which she repeatedly quotes in the film. The gender-fluidity that is expressed through the choice of having Lily speak with a male voice-over troubles her inscribed existence both as a bitch (associating her more intimately with the world of animals) and as a canine philosopher (aligning her more distinctly with the human realm).

One of Derrida's essays is of particular importance to Lily: 'The Animal that Therefore I am (More to Follow)', which he wrote as a contribution to the 1997 Cerisy-la-Salle conference devoted to his work and titled 'The Autobiographical Animal'. This essay has been variously read as a radical plea against the industrialized treatment of animals, including excessive farming and slaughtering on a massive scale. Indeed, his central analogy is the Holocaust. In *Lily Does Derrida* this text functions as the main reference point for the canine philosopher to debate, probe and test the difficult question of 'the' animal, its function, role, meaning and so forth, for humans. Pointing us, in a verbal gesture of self-irony, to High's role in the film as simply provider of images, Lily says: 'She, the she who makes this video for me, she does not know the answers, so she will show you some examples.' Via an extended montage of images viewers are taken to a North American country fair where trained and domesticated animals are showcased in an agricultural context. The sequence includes visual material of cows advertised for their 'dairy-ness', cattle used in a betting game, piglets running down a race track, dogs walking on a tightrope, goats being dragged through an agility course and two roosters picking at the cooked legs of their own kind. Special sound and colour effects in conjunction with visual inserts from Derrida's Cerisy-essay, which are also read out by Lily, distance us from the commercialized spectacle of livestock and agricultural exhibitions.

The cine-essayistic formation of a post-mortem chimeric voice for Lily relies here on a striking temporal a-synchronicity between what is *said* or thought by the fabricated character and what is *shown* in the documentary footage. Visual material of the country fair is also juxtaposed with commentary from Lily, in which she disputes Descartes's view of animals as 'machine-like' or 'orderly'.

Instead, she says, referring us to her own speculative contoured subject status as philosophizing canine with another self-ironic verbal gesture, 'we can be rational – we *can* be. But on our own terms, not necessarily on command'. As listeners to Lily's, aka High's, tale we are constantly encouraged to move back and forth between different speaking positions inscribed in the cine-essayistic text: at times, Lily appears to speak for/with the authority of 'the (almost) human that therefore she is', to misquote Derrida – due to her (speculative) ability to speak, think, argue, create in the human language; at other times, the filmmaker's implied authorial voice/subjectivity becomes more audible in the narration, effecting a slight functional shift within the chimeric voice construction from Lily in her role as a speculative *persona* for the dog to that of an *alter ego* for High.

Following Lily's critique of Descartes's anthropocentrism and her thoughts on dogs as rational beings, we hear her deny this capacity to cats: 'Dogs can be', she argues with the same arrogance that human animals often display towards non-human animal others, 'but not many other animals, and certainly not cats'. Once more, High (with Lily's help) seems to gesture here, tongue-in-cheek, to Derrida. In his Cerisy-essay he recounts a special encounter with his little cat, an unnamed feline, in the bathroom. Caught face-to-face with the critter observing and gazing (back) at him, caused the frontally naked man to feel shame. In the light of Lily's cynomorphic comment on cats as unthinking critters, the philosopher's shame appears to be strangely out of proportion, perhaps even misplaced. Why did Derrida not 'seriously consider an alternative form of engagement', it seems we are encouraged to ask, 'one that risked knowing something more about cats and *how to look back*, perhaps even scientifically, biologically, and therefore also philosophically and intimately' (Haraway 2013: 20) For Lily, the pet dog, philosophy just does not provide any positive knowledge on 'the animal'. She says: 'From Aristotle to Heidegger, from Descartes to Kant, from Levinas to Lacan – great minds, but none of them quite get us. I appreciate that Derrida at least wants to stop calling us "the" animal and begins to think individually.'

Lily functions here as High's mouthpiece, sharing her frustration with Derrida, the astute thinker, who generally knows so well how to avoid the pitfalls of the 'basically imperialist, if generally well-intentioned, move of claiming to see from the point of view of the other' (Haraway 2013: 21). As Haraway writes: 'Derrida correctly criticized two kinds of representations, one set from those who observe real animals and write about them but never meet their gaze, and the other set from those who engage animals only as literary and mythological figures' (21). A possibility for gaining positive knowledge with and of animals is offered, she argues, in the work done by ethologists and animal behavioural scientists such as Gregory Bateson, Jane Goodall, Marc Bekoff or Barbara Smuts. They have met the gaze of living, diverse animals in sustained contact with them, and in response have 'undone and redone themselves and their sciences' (21).

Such doing, undoing and redoing is also suggested in/through *Lily Does Derrida*. The chimeric voice that is fabricated in this cine-essay considers practices of communication that lie outside 'the writing technologies Derrida did know how to talk about' (21). More than a word, High's *animot* moves, walks, thinks and speaks. Lily can no longer look back in a literal sense, as she is going blind and already had to have one eye surgically removed. But through her chimeric voice she can still talk back, that is, intersect the philosopher's voice and make us pay attention to *her* world. The fabrication of a speculative post-mortem voice thereby allows High/the fabulous creature to also 'look back' on her existence *as* dog, even after death. Parts of this existence need to be traced through other-than-human speak.

Lily's general dog-ness and – to a lesser extent – individual dog personality come into focus when one looks at, and also listens to, sequences in which her embodied speak, including facial impressions, bodily gestures and prelinguistic sounds, is amplified. On repeated occasions, the voice-over narration stops and suggests a shift in attention from voice-over to image. At such moments, High's pet dog can simply be seen to move around the house, from one room to another, turn her head around corners in apparent response to her off-screen human guardian or announce herself by the jingle of her dog tag or the repetitive click-clack of her nails scuttling over the laminate floor. In a sequence filmed in High's kitchen and living room area a very low camera positions spectators close to the ground, inviting identification with the assumed perspective of the dog and 'the rooted-in-unglamorous-materiality' (Baker 2013: 18) of the cine-essayistic work. High becomes only briefly visible in this scene in her role as food provider through a pair of legs and an arm that puts down a food bowl. Lily, by comparison, can be seen to weave around High's legs, wag her tail or focus intensely on her human carrying the meat bowl. Short phrases such as 'important food break' over-write parts of this scene and gesture in a humorous way towards High's implied authorial voice.

Although these crawling titles appear to have the same function as a voice-in/off, they reach viewers in written form, as *langue* rather than *parole*. Human speech is muted while the embodied gestures of the live animal are foregrounded.

The documentary footage provides viewers here with what appears to be the daily choreography of well-rehearsed human commands and non-human animal responses. Vicki Hearne in *Animal Happiness* calls such (and other) responses the 'broad talk' of dogs or their ability to speak in a multiplicity of potential 'dialects' (1994: 134), depending on the given situation or the specific human and non-human animal relation formed in a particular situation. Lily's projected dance with her human guardian in the kitchen also reflects in an unusual way Haraway's 'technological polis', based 'on a revolution of social relations in the *oikos*, the household' (1985: 67), which is typical of knowledge production as/in multispecies storytelling.

The (false) impression that we know Lily so well at the end of the film has obviously to do with the sophisticated off-screen narration/voice-over produced for her by the filmmaker. *Lily Does Derrida*, however, also appears to develop an argument about the radically unknown other-ness of animals. Later in the film, High includes images of road kill gleaned from the American country road: close-up shots of a dead cat, a run-over fox and a squirrel – evidently in the process of decomposition. These images are ethically charged, as the gaze of documentary here solicits a cruel and violent intimacy with the non-human animal other. Over these images another set of crawling titles is inserted, which refers us once again to the French philosopher: Derrida asked: 'Does the animal dream?' – another way of asking: Does the animal think? Does the animal produce, represent? Does the animal have a language, and what language? 'Does the animal die? Does it laugh? Does it cry? Does it grieve? Does it get bored? Does it lie? Does it forgive? Does it sing? Does it invent? Does it invent music?' The inverted commas and Thorson's monotonous intonation of these questions signal another form of voice-borrowing; she is now no longer in a strict sense 'speaking' but reading from Derrida's text. The questions he asks, which here become part of the speculative scenario, through the figure of Lily now also seem to constitute High's unresolved questions regarding 'the animal' and her pet dog, in particular. From the vantage point of Lily's post-mortem voice these questions can be answered unanimously: 'Of course, we can and we do,' she responds, tongue-in-cheek, to Derrida's philosophical problem. But this doing/undoing is, as we know, part of a more complex projection made possible through the cinematic medium. Documentary footage of the dead pet dog lying with wet fur on a soiled blanket provides viewers with evidence of Lily's death. But due to the chimeric voice projection High's fabulous creature can live, die, dream, get bored and reinvent herself again and again.

Conclusion

In the construction of trans-species vocal projections humour and irony are often used as rhetorical strategies to make the conditions under which these chimeras can exist as live action animals *and* speak in human languages become audible. In this objective High is certainly not alone. The filmmaker engages through her animal figure with Derridean notions of 'the animal' in the singular, but also provides rhetorical structures that undermine such thinking in an ironic and irreverent way. Addressing us through the speculative post-mortem voice of her canine figure while simultaneously showing us images of her late pet as an at times greedy and pleasing, at other times inarticulate, snoring or shuffling-and-sniffing-around sentient bitch, oblivious to the philosophical discourse, High opens up a gap between voice-over narration and on-screen animal to the effect that 'each deterritorializes the other' (Deleuze and Guattari 1987: 22). The contrapuntal use of what is *said* (on the level of voice) and what is *shown* (on the level of image) underpins logics of dis/identification and absence/presence, by means of which, at times, the anthropomorphized character (the animal we appear to know) and, at other times, the live action animal (the sentient being that remains in the last instance radically unknown to us) is foregrounded.

The chimeric voice with which Lily communicates both in human language and through dialects – which are peculiar to her as a cross-breed street dog or which gain significance in relation to her human guardian, as a pet living with two cats and a woman – does not give us any further insights into the 'consciousness', 'mind' or 'inner life' of the live animal. In creating an anthropomorphized voice for her pet, however, High is able to play on/with a popular human fantasy based on a deep-rooted desire to make animals, particularly companion species, talk back for once and respond to our problems in sophisticated ways. The eponymous Lily represents such a 'creature of imagined possibility' (Baker 2013: 76). In bringing two signifying modes – philosophical discourse and canine speak – together in the imaginative/imaginary space of her cine-essay, the filmmaker makes an attempt at learning something more about/with her dog and at encouraging audiences to reflect more critically (back) upon human-centred notions of what constitutes a 'proper' language, communication, consciousness or existence.

Notes

1 The word 'chimera', as Lynn Margulis and Celeste A. Asikainen argue in *Chimera and Consciousness*, connotes 'questionable, mythical, imaginary' and 'represents the successful merger of two or more distinct beings' (2011: 4).

2 The indefinite article 'a' before 'human language' is used here and throughout this chapter to not give the false impression that the *English* language with which these animals speak is being equated with *human* language.

3 In *Carnal Thoughts* she refers us to a scene from Jean Renoir's *The Rules of the Game* (1939), in which a rabbit is shot dead. The film, as she writes, 'did not merely appropriate a real creature's life for its fiction but also appropriated his death' (2004: 228). This death – which was real, not just a performance by an actor – startled and moved her in a sensual and corporeal way. It opened, as she argues, 'the irreality of the fictional into a different ontological order of representation – namely, into the reality of a documentary space suddenly charged with existential and ethical investment' (270).

4 One of the better episodes (7, series 2) is that in which a baboon and a tree frog are dubbed with Stephen Fry's voice-over. The monkey (as host) and the frog (as player) take part in a quiz that calls up *QI* (Quite Intelligent), BBC's witty panel game quiz show, which Fry hosted for many years.

5 See, among others, '6 Animals Who Can Speak Human' on *Good Mythical Morning*, an online sketch show that can be watched on YouTube.

6 One exception is Andrew Grantham's YouTube sensation *Ultimate Dog Tease* (1 May 2011). In this short film, in which voice artist Grantham dubs a home movie, a canine strikes up a speculative conversation with his human guardian. Human speech and animal speak echo, mimic, intersect and translate one another through word imitation and the frequent use of non-lexical conversational sounds like 'uh', 'yeah' or 'whoah', by means of which a more fluid notion of trans-species communication is re-imagined. These non-verbal expressions emphasize the non-verbal parts of human language and draw attention to the inarticulate aspects of human subjectivity. Grantham's creaky, guttural pronunciation of the dog's words, however, also humorously suggests how laborious it can be for a canine to articulate himself in the human language. The canine both becomes a clever deipnosophist and remains a critter relying on a very different language and communication system from our own. In 2012, due to the popularity of Grantham's YouTube video, shelter dog Clark G, the Belgian Shepherd at the centre of *Ultimate Dog Tease*, was announced official 'spokesdog' of the American Humane Hero Dog Awards™. For alternative dubbings of the same footage, see the YouTube movies *Talking Dog Neutered* (7 February 2012), *Clark G, the Talking Dog Wants an Award*

(16 March 2012) and *Dog Gets Teased about His Superhero, King, the Malinois Military Working Dog* (7 July 2016).

References

Armbruster, K. (2013), 'What Do We Want from Talking Animals? Reflections on Literary Representations of Animal Voices and Minds', in M. DeMello (ed.), *Speaking for Animals: Animal Autobiographical Writing*, 17–33, New York: Routledge.

Asikainen, C. A. and L. Margulis (2011), *Chimeras and Consciousness: Evolution and the Sensory Self*, Cambridge, MA: MIT Press.

Baker, S. (2013), *Artist/Animal*, Minneapolis, MN: University of Minnesota Press.

CBS News (2013), '*Talking' Dogs Say I Love You*, [TV segment] 4 April. Available online: http://www.cbsnews.com/videos/talking-dogs-say-i-love-you/(accessed 25 November 2017).

Clark G, the Talking Dog Wants an Award (2012), [online broadcast] YouTube, 16 March. Available online: https://www.youtube.com/watch?v=_5g6z8uOH2 (accessed 19 January 2018).

Deleuze, J. and F. Guattari (1987), *A Thousand Plateaus: Capitalism and Schizophrenia*, trans. B. Massumi, Minneapolis, MN: University of Minnesota Press.

DeMello, M. (2013), 'Introduction', in M. DeMello (ed.), *Speaking for Animals: Animal Autobiographical Writing*, 1–14, New York: Routledge.

Derrida, J. (2008), *The Animal That Therefore I Am (More to Follow)*, ed. M. L. Mallet, trans. D. Wills, New York: Fordham University Press.

Dog Gets Teased about His Superhero, King, the Malinois Military Working Dog (2016), [online broadcast] YouTube, 7 July. Available online: https://www.youtube.com/watch?v=8fFpY0vXvEM (accessed 10 July 2017).

Good Mythical Morning: '6 Animals Who Can Speak Human' (2014), [online broadcast] YouTube, 3 November. Available online: https://www.youtube.com/watch?v=5ECfyrvD7Kc (accessed 25 November 2017).

Gordon, J. (2010), 'Talking (for, with) Dogs: Science Fiction Breaks a Species Barrier', *Science Fiction Studies*, 37 (3) (November): 456–465.

Grantham, A. (2011), *Ultimate Dog Tease*, [online broadcast] YouTube, 18 November. Available online: https://www.youtube.com/watch?v=nF45ReaFSYs (accessed 5 July 2017).

Haraway, D. (1985), 'A Manifesto for Cyborgs: Science, Technology, and Socialist Feminism in the 1980s', *Socialist Review*, 15 (2) (March–April): 65–107.

Haraway, D. (2013), 'SF: Science Fiction, Speculative Fabulation, String Figures, So Far', *Ada: A Journal of Gender, New Media, and Technology* 3. Available online: http://adanewmedia.org/2013/11/issue3-haraway/(accessed 5 July 2017).

Hearne, V. (1994), *Animal Happiness*, New York: HarperCollins.
Herman, D. (2016), 'Animal Autobiography; or, Narration beyond the Human', *Humanities*, 82 (5) (October): 1–17.
Kuzniar, A. A. (2006), *Melancholia's Dog: Reflections on Our Animal Kinship*, Chicago, IL: University of Chicago Press.
Rudy, K. (2013), 'If We Could Talk to the Animals: On Changing the (Post) Human Subject', in M. DeMello (ed.), *Speaking for Animals: Animal Autobiographical Writing*, 149–159, New York: Routledge.
Sobchack, V. (2004), *Carnal Thoughts*, Berkeley, CA: University of California Press.
Talking Dog Neutered!!! (2012), [online broadcast] YouTube, 7 February. Available online: https://www.youtube.com/watch?v=oJjdBJIrwcc (accessed 10 July 2017).

10

Animating Voices, On-Screen and Off, in Kathleen Shannon's *Working Mothers* (1974–75)

Jean Walton

'Mothers are people,' a young woman insists, as though she shouldn't have to point out the obvious. 'To have a child is just a biological happening, I mean … you don't have to be punished because you just happen to be the sex which bears them.' In the short seven minutes of film that unrolls, we have seen her filling a test tube in a laboratory, then drawing a needle through fabric in what seems to be her domestic space, while her two kids make sandwiches in the kitchen. But mostly she talks, in response to barely audible questions from an off-camera female interviewer sitting near her – talks about how she had to lie to get her job, promising never to let her kids get in the way of her responsibilities; about the 'uncanny' places she's had to leave her children while going to work; about how Canada seems behind her native Jamaica where childcare and youth services are concerned. Identified in the accompanying educational pamphlet only as a 'research biologist' named 'Joy', she is one among a cross-section of women interviewed for *Working Mothers*, a group of films directed by Kathleen Shannon in 1974–75 as part of the National Film Board's (NFB) Challenge for Change programme in Canada.

The project was originally conceived as a fifty-one-minute documentary mandated by one of the Film Board's partners, the Department of Manpower and Immigration, to address the pressing question of why women were not re-entering the workplace after giving birth. But it was recast as a series of shorter films and played a pivotal role both in the Challenge for Change programme and in the inauguration of Studio D, North America's first (and only) state-funded studio devoted exclusively to films by and about women. Shannon (who was to become Studio D's first director) had shot many interviews with individuals and groups of women and was attempting to wrestle them down to a single documentary when NFB distribution consultant Doris Mae Oulton saw the footage and argued that it would work better as a group of short films, with each dialogue or conversation given its own integrity. The shorter films could be shown modularly, in pairs or groups, or however it seemed appropriate for a given audience and used as the basis for discussions and workshops. Thus, according to Rina Fraticelli, 'To counter resistance [based on budgetary concerns] to the proposed new format from both the NFB and its federal sponsor, a team of filmmakers, distribution staff, and community workers were dispatched across the country to conduct a series of test screenings of rough cuts of the films with small groups of women in a range of local community settings' (2010: 305).

This distribution practice was established as a central methodology of Studio D in subsequent years. Although Shannon completed the editing of the interviews as short films, she considered them to be 'unfinished' in the sense that they could only truly be concluded when women discussed them in facilitated environments. Indeed, the films seem deliberately constructed to leave questions unanswered, to pursue avenues that are left for viewers to continue exploring and to convey a combined sense of urgency and hesitancy. Each ends with an unexpected freeze frame – when a speaking woman's face, sometimes in mid-sentence, is suddenly arrested into a grainy still, then fades from the screen. The singularity of one woman's plight is thus shuffled back into the plurality of all women, as still images of other faces from the group of films slowly fade up, one into the next, and the credits appear. This is true during Joy's monologue. 'We *do* know that mothers work and mothers want to work,' she reminds us. But the powers-that-be 'turn a deaf ear, as they've always done to women, especially when societies are changing, they never are anxious to reckon with the voice of women, never!' Freeze frame, and the credits come up, a signal for the 'social animators' who have just screened the film to turn up the lights and invite the audience members to make *their* voices heard.[1]

In the pages that follow, I contextualize the *Working Mothers* group of films in terms of its government sponsorship and its relation to the NFB's Challenge for Change programme. But more importantly, I explore its deployment of voice – as a formal element in the films themselves, as an 'acoustic mirror' in which audiences come to discover their own capacity for articulation and as a specifically political property – the 'voice of women' that a nation-state is reluctant to reckon with. This all but forgotten group of films occupies a liminal moment where state-sponsored cinema for social change intersects with the emergence of feminist consciousness raising; and where the revolutionary 'film action' screening in Latin America's Third Cinema movement was recast as the 'multiplier workshop' in outposts across the Canadian North.

Challenge for change, state sponsorship, mirror vs. hammer

The Challenge for Change programme (1967–80) was piloted in the mid-1960s, with the pioneer director Colin Low's Fogo Island films. Fogo was a small Newfoundland community undergoing immense hardship due to the decline of the fishing industry. The federal government wanted to relocate the inhabitants, more than half of whom had been forced onto welfare, from this remote island to the mainland. With the help of local community leaders, Low shot a series of interviews with residents; then, rather than edit the interviews horizontally into a feature film, he developed what he called a 'vertical editing' system, leaving each interview on its own, to be played back in community situations as the islanders worked out solutions to some of their problems and proposed them to the federal government. The Fogo project was such a success that the method of bringing filmmaking, and then video production, to local communities was established as the modus operandi of the newly formed Challenge for Change programme. This programme's primary goal was to address poverty in Canada, although as time wore on its aims broadened to challenge inequities in the workplace, or problems facing First Nations people, or racism in a Halifax neighbourhood and the like. Each new initiative in the programme was co-sponsored by the NFB itself, and by one or more governmental departments with an interest in eliciting citizen participation in resolving economic and social issues – the Departments of Agriculture, of Labor, and Regional Economic Expansion, or Indian and Northern Affairs, for instance.

The programme had already been up and running for several years when, in the early 1970s, a request came from the federal Department of Manpower and Immigration[2] to fashion a documentary film that would offer a 'definitive answer' to such questions as 'Why were working women not rejoining the labour force after motherhood? What obstacles were they encountering? Why were there wage discrepancies between men and women?' (Fraticelli 2010: 305). Despite the Challenge for Change programme's professed aim to address the needs of Canada's disenfranchised populations, it had failed, up until now, to recognize 'women' as a likely constituency for its initiatives. It wasn't until the prompting of the Department of Manpower and Immigration, pressure from initiatives like the Royal Commission on the Status of Women (whose 1970 report included scores of recommendations for changes to bring about gender equity), and vigorous lobbying from NFB personnel like Kathleen Shannon and her colleagues, that the Challenge for Change programme seemed able to imagine that 'women' might be in need of the kind of media-facilitated political organization that other groups were receiving across Canada (Fraticelli 2010: 304–305).

In proposing to co-sponsor a documentary on working mothers, the Department of Manpower and Immigration had the effect of emphasizing more than one kind of diversity in the workforce – that is, gender diversity (with the aim of maximizing conditions for women in the workplace, even if they had also become mothers) and cultural diversity (reflected through the consolidation of 'immigration' and 'manpower' into the same department). To some extent, the emphasis on cultural diversity extended into Shannon's finished group of films, and played a role, as we will see, in how 'voice' functioned in them – aurally and politically.

What distinguished the *Working Mothers* films – and indeed all of the Challenge for Change projects – from other activist documentary initiatives in North America (and perhaps in the Western world) was the fact that the government itself was footing the bill to mobilize its own citizenry through the deployment of the moving image. Government funding meant, by definition, potential for recuperation of radical efforts by activist filmmakers and videographers if they strayed too far from the interests of the state, and from the nationalizing mission that brought the NFB into existence in the first place (Marchessault 2010: 357). As John Grierson had put it, it was the aim of the NFB to 'interpret Canada to Canadians, and to the rest of the world' (Marchessault 2010: 357); in the case of a nation premised on immigration, this would mean to represent diversity while at the same time promoting unification. If, as Elizabeth Anderson has argued, Studio D (as a division of the NFB) shared in this aim of 'nation-building', it was this very 'reliance on this belief that restricted their project in the end. For, in taking seriously the belief that documentary film could potentially unify the Canadian nation, one must also subscribe to the idea that nations are by definition exclusive and exclusionary' (Anderson 1999: 56).

Where political documentary filmmaking was concerned, two conflicting metaphors recurred in discussions of the mission and efficacy of the NFB and in particular, of the Challenge for Change programme: those of the mirror and of the hammer. It was Grierson who introduced these images, remarking that

> the idea that a mirror held up to nature is not so important in a dynamic and fast changing world as the hammer which shapes it. … It is as a hammer, not a mirror, that I have sought to use the medium that came to my somewhat restive hand. (Grierson 1966: 29)

If the mission of the NFB was to 'interpret Canada to Canadians', then there was always a question of which metaphor would dominate in that process: Did filmmakers seek to hold up 'mirrors' within which Canadians would see themselves? Or would the emphasis be placed more on the process of 'interpretation' – the film as an image-making artefact, a creative 'hammering' out of actuality or the film as an argument that would hammer home a necessity for change? Assessing the *Working Mothers* films in 1976, Dorothy Todd-Henaut and Jennifer Clark revisit these conflicting images and attempt to reconcile them:

> Grierson believed that film should be used as a hammer rather than a mirror, to hammer home a message that expressed the Canadian reality. It isn't as simple as that. The mirror the films have held up to the women in *Working Mothers* helps them become stronger, so they can take up their own hammers and build their own world. Perhaps a good and honest filmmaker can create both a hammer [and] a mirror, and both will have an impact on the course of events. (Todd-Henaut and Clark 1976: 29)

Acoustic mirror

By re-embracing Grierson's rejected mirror, what Todd-Henaut and Clark so succinctly express is the way the cinematic screen replicates the power of the mirror to give us our sense of self, to 'strengthen' our ego development, or as Laura Mulvey puts it, to offer 'structures of fascination strong enough to allow temporary loss of ego while simultaneously reinforcing the ego. The sense of forgetting the world as the ego has subsequently come to perceive it (I forgot who I am and where I was) is nostalgically reminiscent of that pre-subjective moment of image recognition' (Mulvey 2014: 363). In the *Working Mothers* films, the mothers depicted on the screen become points of identification, allowing one to 'forget' for a moment one's relative position of weakness in relation to a public sphere that excludes mothers, and to 'become stronger', or take on the identity of the articulate 'person' who speaks with authority, identifies inequities, points to structural flaws. This is accomplished through the classic structure of the interview, in which 'Shannon plays the role of an engaged listener and witness to each woman's account, standing in for the eventual audience and anticipating the dynamic of the screenings [and creating] a sense of formality that validates the subject's voice and experience within an atmosphere of intimacy and confidence' (Fraticelli 2010: 307). The visual mirroring is thus reinforced by an even more important 'acoustic mirroring', wherein the element of the voice is produced, reproduced, relayed and deployed as the primary tool of social change.

In *The Acoustic Mirror*, Kaja Silverman demonstrates how the female voice is constructed in classic narrative film (and film theory) in relation to the male subject: woman is figured as producer of a maternal voice experienced as a sonorous envelope, as babble, as what must be rejected by a developing child, in order to emerge as an articulate subject in his own right. The child first 'hears' itself in the mother's voice, Silverman writes, refining his 'own' voice through a process of identification, the mother 'lubricating' the 'fit' between child and acoustically mirrored image. But, in a gendered culture, this identification is followed by repudiation, an 'externalizing displacement onto the female subject of what Kristeva would call the "abject" – the vocal and auditory "afterbirth" which threatens to contaminate the order and system of "proper" speech' (Silverman 1988: 80). I would argue that the *Working Mothers* series revalorizes the cinematic function of the maternal voice, using both its gendered quality and its injection of non-mainstream accents, as a mechanism for political mobilization.

Group, individual, accent

Of the eleven films in the collection, three bring together women in groups: *It's Not Enough* (Shannon 1974) served as an introductory film, offering, through voice-over, statistics on wage inequities between women and men, prevalent beliefs that women go to work for 'frivolous' reasons and observations about the resemblance between domestic chores and the occupations available to women in the workplace. Shots of women in an industrial laundry, a cafeteria, a garment factory or at a telephone exchange are interspersed with talking heads of women offering analyses. *... and They Lived Happily Ever After* (Shannon 1975) presents a group of women talking with teenage girls as they express their naïve expectations of married life. *Extensions of the Family* (Shannon 1974) (sometimes left off of the list of films) consists of an interview with a group of families living out an experimental cooperative arrangement, sharing domestic chores and childcare among themselves while also participating in the labour force. An animated film, *The Spring and Fall of Nina Polanski* (Hutton and Roy 1974), spins out the tale of a newly married housewife who, little by little, turns into the automated machines that define her household, until she escapes into nature. The remaining seven films feature a single subject whose spoken monologues are drawn out by Shannon's barely audible off-camera questions; it is these films that most exemplify the deceptively simple technique Shannon has used to deploy voice as a validated (rather than repudiated) acoustic mirroring element and to extend it into the workshops that followed the screenings.

Within the films themselves, Shannon chose to represent women across a range of cultural backgrounds. Thus, five out of the seven individuals she interviews speak from outside a white, mainstream context. In *Tiger on a Tight Leash* (Shannon 1974), a university professor speaks with marked British accent; in *Luckily I Need Little Sleep* (Shannon 1974), a nurse who has emigrated from Greece addresses us with breathy Greek inflections; Joy, in *Mothers Are People* (Shannon 1974), is obviously from Jamaica – her background marked as much by her accented voice as by her skin colour. In two other cases, the women represented are of First Nations origin; one, a Métis woman of Cree background (in *Like the Trees* [Shannon 1974]) and another, a member of the Abenaki people (*Our Dear Sisters* [Shannon 1974]). The singularity of position from which each woman speaks is signalled first and foremost through her voice – the inflection of her English as not mainstream 'Canadian' – that is, not coinciding with the CBC broadcast voice that had hitherto dominated the airwaves of English-language Canadian media. Shannon's infrequent questions, posed in her often-hesitant West Coast Canadian cadence, serve by contrast to emphasize both the assertiveness of the others and the fact that they speak from a decided 'elsewhere' to 1970s Canada.

In many cases, the non-mainstream voices are the vehicles for utterances whose content juxtaposes undesirable conditions in Canada with more favourable conditions for women elsewhere: in the opening of *Mothers Are People*, for instance, Joy contrasts the progressive services available in Jamaica (purportedly a 'developing nation') to Canada's empty 'lip service'. In *Luckily I Need Little Sleep*, Kathy expresses a preference for the Greek practice of retaining one's maiden name as one's professional name, in contrast to her use of her husband's name in Canada. She also notes that in Greece, day-care centres are more available than in Canada, even in less populated areas. Even if a Canadian policy is not explicitly contrasted with that of another country, the accent signals a difference from which the interviewed subject speaks. In her clipped British accent, for example, Cathy, in *Tiger on a Tight Leash*, speculates on why day-care centres are 'so slow to come about' in Canada: 'Who's going to pay for them, and whose responsibility is it to start them?' she begins. The working mother hasn't the time, energy, political clout or money to establish broadly available childcare:

> We just haven't reached the stage where it's widely accepted that it is somebody else's responsibility – or the national interest is not yet such that it becomes the job of politicians and organizers of our social structures to provide these things (freeze frame).

The inflection of her British accent implies an 'elsewhere' from which she analyses the Canadian National Interest.

For Rose, a Métis in Alberta, the critique of national Canadian culture is not so much about its lack of services for working mothers, as the way it has alienated First Nations women like herself who have encountered discrimination when trying to eke out a living within the mainstream workforce. Fed up with dressing, acting and speaking in a certain way, while not making enough money to support her children (who had been left in the care of her family of origin) she gets to the point of dropping out altogether from White Canada, and returns to Cree culture in the rural Woodlands, where she finds a more complete way of life.

The recorded voice within the films, then, is used as the vehicle through which a critique will be launched – from a gendered and non-mainstream Canadian locus. But precisely because these films are left 'unfinished', to allow for their instrumentality as catalysts for discussion, this recorded voice is deliberately truncated, leaving conceptual and temporal space for another voice to emerge. This is where the group discussions come in, but also where the *Working Mothers* project departs in an interesting way from similar film activism in the Latin American Third Cinema movement.

Animating women as community

In 1975, after they had been showing the films around Canada, Shannon and three of her fellow social animators recorded a conversation among themselves about how their directed workshops had been going. The finished transcript was published in the next issue of the in-house Challenge for Change newsletter and served as a training guide for new generations of animators. The women were thus not only recounting their experiences in the animation sessions, but working inductively towards some shared understanding of how best to use one's own 'trained' voice as a mediating tool between the mechanically reproduced voices on screen and the tentative voices of the women in the audience.

In this way, the *Working Mothers* group of films, and the meetings and workshops in which they were shown, were redeploying the 'acoustic mirror' *as* 'maternal voice' that had become so formulaic in classic cinema, emphasizing its role not as incomprehensible babble (as experienced by the infant), but as facilitator of entry into language, specifically for a female subject this time. Shannon and her fellow social animators take up the role of the mother, introducing their audience members to the 'mirror reflection' on the screen, 'lubricating the fit', and encouraging the coming into being of the speech of their audience members. Moreover, it was not only women or mothers as individuals who were addressed, but also women as a political group, linked in their commonality as actual or potential mothers. If the obvious had to be asserted, that *Mothers Are People*, then it was because for so long, the role of the mother had been only to perform that function of facilitating the male subject to take up his position in the domain that defines what is human: that is, the exercise of language within a signifying system. Or in other words, in the public sphere of 'work'. But there remained the challenge of how to aggregate women *as* a community. This was accomplished through the revalorization of Grierson's rejected mirror image and its deployment in the context of social animation.

If it was more or less Colin Low's vertical editing system that was adopted by Shannon for her *Working Mothers* series, the question of where to find the 'community' in need of intervention was not as obvious as it was among Newfoundland fishing villages. 'Working mothers' were not so easily defined by locality, but were dispersed all over Canada and belonged to diverse communities defined by other regional, ethnic and class factors; it could not be taken for granted, therefore, that they identified already as a community themselves. Community among working mothers was thus the goal, rather than the starting place, for film-based political intervention.

Moreover, when women gathered to see the films, the NFB animators could not themselves seem to offer an already politicized collectivity that one could simply 'join'. Or as Shannon had put it:

> People who come to see the films expecting to be told 'where the women's movement is at' will be disappointed. The women's movement is wherever a group of women meet together as women. [The films are] used as ways to get people communicating. … The people coming together are the event and the films become a contribution to that event. (Quoted in Fraticelli 2010: 306)

The taped conversation among the *Working Mothers* animators tells us a lot about how the facilitation of a 'political voice' among women as a 'community' of 'mothers' was shaping up. Shannon opened the discussion by asking, 'What is the difference between a workshop and a screening? What are the qualities of a good animator? I'm hoping this issue [of the Canadian Film Centre (CFC) newsletter] will help animators be a little bit better, whatever stage they're at' (Shannon et al. 2010: 53).

What emerges from the discussion is a conflict between the animators' obvious desire, on the one hand, to engender in their women audience members a recognition of the structural conditions that underlie their oppression and frustration (lack of government-sponsored day care, absence of programmes to allow temporary part-time work schedules, widespread trivialization of women's labour etc.) and on the other hand, to create a space for these same women to speak for themselves, and come to their own conclusions about the films, and about what was needed to ameliorate their own situations. As Elizabeth Prinn put it, 'You can't school yourself to appear open. People "hear" you. So maybe with these films we can create a situation which is in a good place, in between a propaganda situation and a "sensitivity" situation' (Shannon et al. 2010: 59).

This practised suppression of one's own experience as a tactic to invite audience members to develop and wield their own voices was part of the Challenge for Change's general ethos of democratization. In fact, not long before the *Working Mothers*' team developed its own particular version of it, Colin Low had a confrontation with John Grierson about this very aspect of the programme. Recounting a class in which he had shown Fogo footage to the documentary pioneer, and tried to explain that 'giving a voice to people' was what it was all about, he recalled that Grierson railed against the passive stance the professional animator was forced to take:

> So the filmmaker is nothing but a tool, a camera operator or projectionist in the formulation of these problems and solutions. What about the intelligence, world experience, expensive education that could be brought to these people? If you have no opinions, no ideas, no commitments, nothing to say, why further burden

> these poor folks, whose lives are difficult enough, with manipulative nonsense? They need all their working energy and time to simply get by. (Low 2010: 20)

So how did one negotiate the fine line between exercising sensitivity in leading a discussion and engaging in 'manipulative nonsense'? Shannon and her fellow animators were well aware of the paradoxes involved, and the likelihood that CFC films might be less democratic than they had hoped, as is clear by the terms in which they described their challenges among themselves, as they debate the difficulty of having to 'choose between laying my trip on other people or putting a padlock on my mouth' (Shannon et al. 2010: 54).

Good animators must 'listen with a capital L', they observed, 'responding to what they hear' (Shannon et al. 2010: 54); the animators' 'opinion isn't the important thing'; rather, one must 'draw people out' so that by the end 'everybody is talking' (Shannon et al. 2010: 54). The unfinished films seem in some way to model what the animators understand themselves to be doing – as live 'audio-visual' mediators whose sole role is to create the best conditions from which the women in the audience can use their voices to articulate the problems, and possible solutions, that are specific to their local context. None of the women interviewed on-screen, for instance, are identified by name – either through subtitles, the interviewer's voice or in the final credits.[3] Their anonymity is meant to imply that their stories could be anyone's.

Correspondingly, the facilitators seek to efface their own opinions, perspectives and judgements: 'In just the same way as the films don't tell you what to think, the animator can't tell you what to think either,' Shannon comments (Shannon et al. 2010: 55). Indeed, an ethos of neutrality and impersonality is promoted as a skill that involves careful dissimulation: 'Refrain from feeling that there are right answers which you know ahead of time,' says one animator, and another agrees, 'you have to really mean it' (Shannon et al. 2010: 55). This causes some laughter among them, as though in recognition of the irony that they must pretend to be less knowledgeable than they really are. If they don't strike just the right note, and are not entirely convincing in their stance of neutrality, 'that really projects', they say. 'If what you're doing is trying to get a group to come up with an answer you've already perceived, that will colour what happens with the group' (Shannon et al. 2010: 55). The animator should withhold her judgement: 'If the animator relates to the films in judgmental terms, it sets the scene for relation to each other in judgmental terms' (Shannon et al. 2010: 57). Indeed, it is suggested that in training sessions, they 'say something to the effect that people who feel they know what's right for the world and how to change it should disqualify themselves as animators' (Shannon et al. 2010: 57).

Most interesting is when the discussion seems to touch on the very disposition of the animators' bodies as vehicles for facilitation. One must know how to position one's body, standing up so that the group has someone to address, rather than sitting down, which causes the collectivity to disintegrate into smaller groups; one should appear 'physically relaxed', which can only be achieved by actually 'being' relaxed. 'That can be helped', it is added, 'by not having a stake in a particular outcome' (Shannon et al. 2010: 58). In other words, an animator must train herself not to care whether the audience comes around to sharing the insight she has come to have through her own education and experience. It is almost as though the best animator was the one who could convince not only her audience, but herself, that she was no better equipped than they to make sense of the political conditions working mothers found themselves confronting. In a way, the challenge was to make the woman's image and voice, on-screen and off, into a mirroring mechanism rather than a hammering device, so that articulations of problems and solutions would seem to arise from the collective voices of viewers in conversation with each other during the post-screening discussions. Only thus would a truly democratically defined feminist community result from the encounter, a community comprised of subjects who were at once the beneficiaries of mirroring and the agents hammering out political change.

Compañero relator vs. embodied animator

In closing, it is instructive to compare what the *Working Mothers*' animators were doing to similar practices transpiring in South America, where Fernando Solanas and Octavio Getino had been showing their multi-part, extravaganza of a revolutionary film, *Hour of the Furnaces* (Solanas and Getino 1968) in clandestine meetings across Argentina.[4] 'Every comrade who attended such showings', as Solanas and Getino put it, 'did so with full awareness that he was infringing the System's laws and exposing his personal security to eventual repression' (Solanas and Getino 2014a: 247). No longer a mere spectator, an audience member was 'an actor, a more important protagonist than those who appeared in the films' (Solanas and Getino 2014a: 247). In their groundbreaking manifesto, 'Towards a Third Cinema', Solanas and Getino described a developing practice that paralleled in many ways the strategies employed by the *Working Mothers*' animators, right down to the necessity to 'disinhibit' the participants involved. The film itself would 'be much more effective' if it '*sought its own liberation in its subordination to and insertion in others, the principal protagonists of life*' (Solanas and Getino 2014a: 248. Emphasis in original).

In Solanas and Getino's manifesto, the emphasis is placed either on the traits of the new filmmaker (inevitably figured as a man who must subordinate himself to a guerrilla-style collective), or on the film (to be subordinated to the needs of the [male] 'actors' whose militant activity is mobilized by it) or on these spectator-turned-actors themselves, as they go through the liberating process of politicization. But nowhere in their 1969 manifesto, or in their follow-up statements (Solanas and Getino 2014b), do they reflect on the desirable qualities of the leader of post-screening discussions. Indeed, it is almost as though 'Towards a Third Cinema' performs a disavowal of the very existence of this important role in the distribution and exhibition process, by locating the agency for 'animating' audiences in the films themselves – in those passages in the second and third parts of *Hour of the Furnaces*, for instance, in which all images cease, and over black leader, a male voice directly addresses the audience, inviting them as 'comrades' to engage in a post-screening discussion, after which an instruction to the projectionist appears: 'Espacio para intervencion del compañero relator'.[5]

This indexical gesture of the film – pointing to a temporal disjunction where events of the current moment will be 'narrated' to the participants by their *compañero* – is about as close as Solanas and Getino get to an acknowledgement of the labour to be done by someone who is not the filmmaker: this is the person who will be responsible for 'lubricating' the 'fit' between the moving images that are to be projected and the 'actors' who are invited to identify, listen, debate and carry forward revolutionary history itself.

It is hardly worth mentioning the obvious ways the two movements are gendered – Third Cinema with its continual theme of the fashioning of the 'New Man' contrasting sharply with the *Working Mothers*' strategies of self-effacement, for better or for worse. But in the published materials that accompany each of these inaugural moments in the history of political, activist or militant cinema, what I am struck by is the absence on the one hand, and presence on the other, of some sustained attention given to the role of the organizer or animator in what was arguably the most important aspect of this practice of a new cinema: its insertion, through discussion, debate or dialogue, into the emergence of a political voice, and therefore, of political action. Shannon and her colleagues thought it essential to describe and establish a methodology for regulating the very embodiment of the animator's voice, its tone, its seeming openness, its neutrality, its level of relaxation, even the location in the room from which it emanates – it seemed relevant to specify all of this as part of the 'revolutionary' process. In other words, what seems missing from the Third Cinema documents is precisely how the 'animator' functions as part of the cinematic apparatus itself as it extends into the field of revolutionary action. In this respect, perhaps, taking account of the way a given film-action reproduces or defies First Cinema's conventions of the 'acoustic mirror' is one of the most valuable contributions made by the *Working Mothers* chapter in the history of political cinema.

Notes

1 Social animation was a form of community organizing developed by Michel Blondin in Quebec in the 1960s. Translated from the French *animateur social*, the 'social animator' was akin to what would be termed a community developer in the United States. In the Challenge for Change programme, for instance, it was routine for teams of locally hired 'social animators' to accompany filmmakers and videographers in their efforts to bring about political and social change through the moving image. See Druick (2010).

2 The Department of Manpower and Immigration was a relatively new division of the federal government, the result of a restructuring of departments under the Liberal leadership that had taken power in the 1960s. Prior to this, the functions of citizenship and immigration had been handled by one department, and manpower and labour by another. The consolidation of immigration and manpower functions was part of the Liberal government's broader efforts to redress what had been a blatantly racist immigration policy and to fund employment initiatives in such a way as to insure that immigrants, regardless of their country of origin (i.e. whether it was white European or not) would be integrated into the Canadian labor force. See Parai (1975).

3 Only in printed educational pamphlets, or on the VCR or DVD packaging, are identifying first names supplied.

4 In this case, of course, far from being funded by a government agency, this insurrectionary film, with its compelling and explicit critique of the neo-colonialist military regime ruling Argentina at the time, was outright banned.

5 Space for the intervention of our narrator.

References

...and They Lived Happily Ever After (1975), [Film] *Working Mothers* Group, Challenge for Change Programme, I. Angelico and A. Henderson (Dirs), Canada: NFB.

Anderson, E. (1999), 'Studio D's Imagined Community: From Development (1974) to Realignment (1986–1990)', in K. Armatage, K. Banning, B. Longfellow, and J. Marchessault (eds), *Gendering the Nation: Canadian Women's Cinema*, 41–61, Toronto: University of Toronto Press.

Druick, Z. (2010), 'Meeting at the Poverty Line: Government Policy, Social Work, and Media Activism in the Challenge for Change Program', in T. Waugh, M. B. Baker, and E. Winton (eds), *Challenge for Change: Activist Documentary at the National Film Board of Canada*, 337–353, Montreal: McGill-Queen's University Press.

Extensions of the Family (1974), [Film] *Working Mothers* Group, Challenge for Change Programme, K. Shannon (Dir.), Canada: NFB.

Fraticelli, R. (2010), '"Would I Ever Like to Work": The "Working Mothers" Films and the Construction of Community', in T. Waugh, M. B. Baker, and E. Winton (eds), *Challenge for Change: Activist Documentary at the National Film Board of Canada*, 303–313, Montreal: McGill-Queen's University Press.

Grierson, J. (1966), *Grierson on Documentary*, ed. F. Hardy, Berkeley, CA: University of California Press.

Hour of the Furnaces (1968), [Film] F. Solanas and O. Getino (Dirs), Argentina: Groupo Cine Liberacion.

It's Not Enough (1974), [Film] *Working Mothers* Group, Challenge for Change Programme, K. Shannon (Dir.), Canada: NFB.

Like the Trees (1974), [Film] *Working Mothers* Group, Challenge for Change Programme, K. Shannon (Dir.), Canada: NFB.

Low, C. (2010), 'Grierson and Challenge for Change (1984)', in T. Waugh, M. B. Baker, and E. Winton (eds), *Challenge for Change: Activist Documentary at the National Film Board of Canada*, 16–23, Montreal: McGill-Queen's University Press.

Luckily I Need Little Sleep (1974), [Film] *Working Mothers* Group, Challenge for Change Programme, K. Shannon (Dir.), Canada: NFB.

Marchessault, J. (2010), 'Amateur Video and the Challenge for Change', in T. Waugh, M. B. Baker, and E. Winton (eds), *Challenge for Change: Activist Documentary at the National Film Board of Canada*, 354–365, Montreal: McGill-Queen's University Press.

Mothers Are People (1974), [Film] *Working Mothers* Group, Challenge for Change Programme, K. Shannon (Dir.), Canada: NFB.

Mulvey, L. (2014) 'Visual Pleasure and Narrative Cinema', in S. MacKenzie (ed.), *Film Manifestos and Global Cinema Cultures: A Critical Anthology*, 359–369, Berkeley, CA: University of California Press.

Our Dear Sisters (1974), [Film] *Working Mothers* Group, Challenge for Change Programme, K. Shannon (Dir.), Canada: NFB.

Parai, L. (1975), 'Canada's Immigration Policy, 1962–74', *The International Migration Review*, 9(4), 449–477.

Shannon, K., E. Prinn, D. M. Oulton, and I. Angelico (2010), 'Working with Film: Experiences with a Group of Films about Working Mothers', in T. Waugh, M. B. Baker, and E. Winton (eds), *Challenge for Change: Activist Documentary at the National Film Board of Canada*, 52–60, Montreal: McGill-Queen's University Press.

Silverman, K. (1988), *The Acoustic Mirror: The Female Voice in Psychoanalysis and Cinema*, Bloomington, IN: Indiana University Press.

Solanas, F. and O. Getino (2014a), 'Towards a Third Cinema: Notes and Experiences for the Development of a Cinema of Liberation in the Third World', in S. MacKenzie (ed.), *Film Manifestos and Global Cinema Cultures: A Critical Anthology*, 230–249, Berkeley, CA: University of California Press.

Solanas, F. and O. Getino (2014b), 'Militant Cinema: An Internal Category of Third Cinema', in S. MacKenzie (ed.), *Film Manifestos and Global Cinema Cultures: A Critical Anthology*, 256–257, Berkeley, CA: University of California Press.

The Spring and Fall of Nina Polanski (1974), [Film] *Working Mothers* Group, Challenge for Change, J. Hutton and L. Roy (Dirs), Canada: NFB.

They Appreciate You More (1974), [Film] *Working Mothers* Group, Challenge for Change Programme, K. Shannon (Dir.), Canada: NFB.

Tiger on a Tight Leash (1974), [Film] *Working Mothers* Group, Challenge for Change Programme, K. Shannon (Dir.), Canada: NFB.

Todd-Henaut, D. and J. Clark (1976), 'The Challenge for Change Experience', *Journal of Architectural Education*, 29 (3): 28–29.

Would I Ever Like to Work (1974), [Film] *Working Mothers* Group, Challenge for Change Programme, K. Shannon (Dir.), Canada: NFB.

Section Four

Documentary Modes and Techniques

11

The Voice in Documentary Sound Design: A Digital Revolution

Chris Cagle

Documentary scholars frequently analyse broad sonic choices, such as the sound–image relationship, but as a whole they have paid far less attention to sound professionals' contribution to documentary film. As Robert Strachan and Marion Leonard note, 'Discussion of documentary films has often concentrated on the visual content, editing and directorial decision making' (2014: 166). This oversight stands in contrast to the renewed attention to the history and aesthetics of sound design in the fiction film.[1] The difference in scholarly emphasis may spring in part from the often-separate traditions of fiction and documentary production practices. Jeffrey Ruoff, for example, sees documentary sound as working according to fundamentally different principles than fiction sound. 'One of the major stylistic characteristics of documentaries that use sounds recorded on location', he observes, 'is the lack of clarity of the sound track. Ambient sounds compete with dialogue in ways commonly deemed unacceptable in conventional Hollywood practice' (1992: 221). While fiction film shoots and some documentary interviews can control their sonic environments to minimize stray noise, location documentary shooting must work reactively and manage existing unclean sound. In this sense, it is, in Richard Leacock's formulation, 'uncontrolled cinema' (1961). Sound recording is important for fiction, too, but the difficulty of capturing location sound gives the recordist a paramount role in documentary.

Much has changed since Ruoff's 1992 essay, however. Ruoff observed with justification that 'Hollywood's increasing reliance on multi-track post-production techniques contrasts significantly with documentaries that use only location-recorded sound' (1992: 217), but digital editing tools have since made post-production work, including multi-track re-mixing for aesthetic ends, standard for most theatrical and broadcast documentaries. One example of a film without obvious sound manipulation, the character-driven documentary *Queen of Versailles* (Lauren Greenfield 2012), used six post-production crew for Foley or sound effects alone, and its sound editing supervisor ('sound designer'), Peter Albrechtsen, is known not only for his sound design of documentaries such as *Putin's Kiss* (Lise Birk Pedersen 2011) but also for sound effects editing on European feature films, including *The Girl with the Dragon Tattoo* (Niels Arden Oplev 2009) and *Antichrist* (Lars von Trier 2009). Documentary studies has long examined documentary cinema as a construction as much (or nearly as much) as fiction film. This chapter makes a related but more specific claim, namely that over the last two decades documentary sound design has converged with fiction film sound design in its production practices and sonic aesthetics.

Historically, documentary sound lagged behind fiction film sound practice by continuing analogue sound design while fiction turned to digital. Moreover, documentaries typically minimized background sound to give priority to the voice. More recently, digital sound practice and Dolby sound technology have allowed documentary filmmakers to embrace the very environmental elements that earlier generations of documentarians tried to minimize. Strachan and Leonard, for instance, point to recent landscape documentaries that embrace the sonic environment as an integral part of the subject, not merely as background. Their examples, *Silence* (Pat Collins 2010) and *Sleep Furiously* (Gideon Koppel 2008), suggest the fruit of a detailed history of film sound as a craft. Nonetheless, shifting the focus to environmental sound, however fitting for certain films, obscures the importance of the actual human voice in contemporary documentary sound design. While some documentaries, like *Homo Sapiens* (Nikolas Geyrhalter 2016), are without human voice altogether, most grant a central role to speaking subjects. Digital sound design has increased the expressive use of environmental sound in recent documentary, but rather than replace the voice it has renegotiated the relation between environment and voice.

Documentary filmmakers and sound designers can use a number of strategies to balance voice and environment, but I propose that three paradigms define the key approaches used in documentaries made in recent years. The first represents a reconciliation of voice to its environment, which creates a *spatialized voice*. Two films illustrate this spatialized voice well: *Last Train Home* (Lixin Fan 2009), a character-based documentary of a Chinese family's migration between their home village in Sichuan and work in Guangzhou, and *Roundabout in My Head* (Hassen Ferhani 2015), a poetic portrait of a slaughterhouse in Algiers. Working against this tradition, some films exaggerate the human voice's relation to a realist space to create what I am calling *spectral voice*. A good example is *Man on Wire* (James Marsh 2008), an account of Philippe Petit's 1974 stunt walk between the World Trade Center buildings in New York City. Finally, experiments in separating voice entirely from realist space create an *uncanny voice*, exemplified by Vitaly Mansky's *Under the Sun* (2015), a film about North Korea's propaganda. While some documentaries may exhibit other approaches, I argue these three sonic categories reveal the main strategies that documentaries use to reconcile the centrality of the human voice for non-fiction meaning with the environmental sound so important to the overall sound design of many fiction films.

The spatialized voice

In his influential essay 'Narrative Space', Stephen Heath asks, 'What, then, is this "taking place" in cinema?' (1982: 24). For Heath, the answer is that cinema overcomes the static nature of the frame and the two-dimensional screen through editing and camera movement in order to reinforce the ideology of the powerful, seeing individual. Sound, too, plays a role in the construction of space; post-1970s fiction film sound design in particular heightens the role of effects and environmental sound to foster the spectatorial sense of space. However, as Michel Chion notes, the matter is more complicated. 'In actual movies, for real spectators', he writes, 'there are not all the sounds including the human voice. There are voices, and then everything else' (1999: 5). Chion terms this tendency vococentrism, and much like the alternation that Heath describes visually, it means that lusher, environmentally oriented sound design alternates between the space of human voices in dialogue and the space of narrative action and the environment. Moreover, the soundtrack does so in a way that seems integrated for the spectator-auditor. To take an example from the historical drama *Frost/Nixon* (Ron Howard 2008), as reporter David Frost arrives at an airport, a series of effects and environmental noises, including a phone hanging up in a booth, an airplane taking off and an announcement over a public address system, hold forth prominently in the sound mix until the characters' voices quiet them to a considerably lower level. While 'transparency' is an imprecise term for this approach, narrative clarity drives mixing levels, and the shifts in sonic perspective are not immediately noticeable to the spectator.

Documentary sound in the analogue period – lasting from the introduction of mobile magnetic tape recording in the 1960s to its replacement by digital recording technology the 1990s – emphasized capturing the voice with as much clarity as possible and mixing primarily to achieve vocal clarity in the final film. Since then, digital post-production has encouraged extensive re-recording and robust sound mixing along the lines of fiction cinema. In practice, many documentaries occupy an in-between space, using many of the tools of post-production while making use of the imperfections of captured sound. *Last Train Home* combines the immediacy of small-crew shooting with the multi-layered mixing capability of post-production sound mixing. The director, Lixin Fan, was previously a sound recorder, working on another documentary successful on the festival circuit, *Up the Yangtze* (Yung Chang 2007). As director, Fan brought an attention to sound design to *Last Train Home*. Shot over several years, the film combines small-crew shooting with a more extensive post-production mixing. As such, it exemplifies a common fiction filmmaking practice with character-driven documentary.

Last Train Home uses space to allegorize the difference between rural and urban China. The film follows the experience of a married couple Changhua and Sugin Zhang, who both work at a garment factory in Guangzhou, miles from their family in Sichuan. Like other character-driven documentaries, the film uses their individual story to explore the wider social context, in this case the issues generated by the Chinese economic miracle. As the Zhangs pursue economic security, they bear a heavy personal cost to their family life. The geographical alternation between Guangzhou and Sichuan province structures the film thematically and provides a narrative conceit, in which the Zhangs travel on the annual New Year's holiday between the two locations. As the film proceeds, the surface narrative question – Can the Zhangs get on an oversold train to make it back in time? – gives way to the emotional conflict between the Zhangs and their estranged daughter.

Sonically, the film constructs three types of space to correspond to Guangzhou, Sichuan and the train in between. After the prologue, the film's opening proceeds with a series of scenes establishing Guangzhou, the garment factory and the Zhangs' daily routine. There is little recognizable speech, other than Sugin's comment to her husband that he need not help with the chores. The rest of the soundtrack selectively uses environmental sounds: traffic passing by, boxes being unloaded, sewing machines. The mix isolates these in a stereo soundscape that gives the illusion of a complete cinematic space. After these initial scenes, the seeming cacophony gives way to actual competition between sounds. As the Zhangs discuss getting train tickets with their co-workers, the Zhangs' voices are clearly audible, but so are the voices of the others eating with them and those at other tables. Later, alone, the Zhangs talk about their trip home while they work at their sewing machines in a factory. Despite very clear sound recording, the ambient noise of the factory competes with their dialogue, and mechanical sounds threaten to overwhelm their voices. The sound mix segues both the opening and closing of the scene by blending similar ambient noises of the preceding and following scenes. These scenes set up a pattern in *Last Train Home*: environmental sound constructs the space, then the mixing minimizes the recording difficulties in overall 'close-up' treatment of the dialogue, and finally the scene ends with a sonic transition or a return to environmental sound.

This balance comes under strain during the train scenes. Here, the mix departs from the chaotic background noise, and the Zhangs' frantic calls to one another are heard as they rush to get on a train, stow their luggage and grab their seat. Radio microphones help capture their speech, and the mix contrasts the relative flatness of their voices with the more voluminous, if unclear, space of the train. While the earlier scenes at the factory align the Zhangs' voices with their co-workers, implicitly suggesting camaraderie, the train scenes exploit the difference in microphone types and sound perspective to isolate the couple from other passengers and from the various voices of authority, personified in the multiple public address announcements.

During scenes in which their daughter Qin goes about her daily agricultural life in Sichuan, in contrast, the sound design stresses the character's integration into her space. Rather than try to manage the external or ambient sound, Qin's introductory scene exploits it for its possibility of spatial depth. The first shot starts with the sounds of birds, insects and trickling water high in the mix, even before Qin enters the frame, walking in the distance. The next shot is similar, but with more insect sounds in close-up and with Qin's footsteps now audible. With each shot, the sounds of Qin working come to dominate more of the mix, culminating in a shot of her chopping greens while ducks walk in the background and while Qin, in voice-over, gives her testimonial interview. Michel Chion has coined the term 'superfield' to refer to the sonic invocation of offscreen space complementary to the primary sound of on-screen space. The superfield, he states is 'the space created, in multitrack films, by ambient natural sounds, city noises, music, and all sorts of rustlings that surround the visual space and that can issue from loudspeakers outside the physical boundaries of the screen' (1994: 150). Qin's voice comes in the flow of environmental sounds, and as she chops greens her voice (recorded with a directional rather than a radio microphone) is punctuated with these noises, including roosters crowing, ducks quacking and a stream running. Common to fiction films, the superfield illusionistically evokes a larger space of the action, and *Last Train Home* frequently borrows from this practice, particularly in its juxtaposition of rural and urban spaces.

Last Train Home shows how a character-driven documentary can integrate theme and a character portrait with an aestheticized depiction of the landscape. However, other documentaries adopt qualities seen in contemplative art cinema, or slow cinema, exemplified by auteurs like Apichatpong Weerasethakul and Lisandro Alonso. Leonard and Strachan argue that a certain type of poetic landscape documentary upends the normal hierarchy of environmental sound. These films, they note, 'employ a number of editing conventions that accentuate the aural through the manipulation of rhythm and pace. Most striking is the predominance of ambient sound in terms of screen time' (2014: 170). Another way to put their formulation is that the aesthetics of slow cinema, with its predominance of static shots, long takes, 'dead time' and ambient sound (Flanagan 2008), influence many poetic documentaries.

Roundabout in My Head is a variation of this type of poetic documentary. Filming the spaces of an Algiers slaughterhouse rather than a landscape per se, it has relatively little dialogue; there are some recorded conversations and some testimony addressed to the filmmaker. Much of the film consists of long takes of the building and its workers, with a heavy use of environmental noise. Director Hassen Ferhani (2016) has even noted that he was drawn to the space for its rich array of sounds. The film opens with a shot of the main subject, Yusuf, operating a pulley. The shot holds as the squeaking sound reverberates off the plaster walls in the cavernous space of the room; in the background a number of sounds of the slaughterhouse echo in the mix. The film maintains this approach over much of the film, confirming the aesthetic that Strachan and Leonard identify: 'Rather than being a background which naturalises the audiovisual meld, or providing a bed for foregrounded sound, ambient sound here speaks with a powerful voice' (2014: 177). Like similar contemporary documentaries, *Roundabout in My Head*'s soundtrack is the result of extensive mixing and rerecording. It includes Foley sound, as in the close-up sound of a worker hacking at a cow's leg bone (the shot is in long shot, away from any microphone). Moreover, it mixes ambient sound, Foley and recorded dialogue with an exaggerated dynamic range.

Even in a film this rich in its environmental sound recording, the voice still plays a central role. Ferhani punctuates the static shots of the space with interviews with a few workers, focusing particularly on Yusuf. These men talk about daily life but also the failed promises of the Algerian state. Since *Roundabout* does not provide a broader exposition of Algerian politics, the implications of these interviews can be difficult to parse. Nonetheless, the poetic structure invokes what Oskar Negt and Alexander Kluge (1993) would term an 'alternative public sphere', a space of political deliberation separate from the official public sphere and based in the lived experience of the working class.

The sonic recording and mixing emphasizes the communal experience of the workplace space. Sound mixers frequently use what they call walla, the indistinguishable talking added to give a sense of people interacting in the background without competing with the primary spoken dialogue. *Roundabout* does not use walla in a true sense but its recorded voices serve a similar function, particularly as their echo renders them indistinct. There is constant activity, moreover, implied by the soundtrack. Machinery, trucks, animal sounds and water all have a distinctly echoed and distant quality to them, and various human voices, particularly those of the workers but also those of recorded songs, fill the background with a similar sonic rendering. In *Roundabout*, voice becomes environment, part of the rich texture of the subjects' lives. Even the conversations with the subjects combine with environmental sounds in a manner different from *Last Train Home*'s attempt to minimize the ambient. In an outdoor conversation with Yusuf and a man identified only as The Kabyle, echo-heavy slaughterhouse sounds form the backdrop of the men's conversation while raindrops are very loud in the mix. This effect could be motivated as a realist capturing of the sounds closest to the microphone, but the men have wireless radio mics, which would not pick up the raindrops. Later, near the film's end, a homeless man walks through a courtyard or square, and his sounds are in close-up, in contrast with the sonic depiction of this space earlier as a cacophony of distant, echoed sounds.

Much like *Roundabout*'s crisp high-definition digital videography, the sound of its subject's speech tends to be very clear, the radio microphones and augmented place in the mix adding to a type of hyperrealist space. This film and many other poetic documentaries exaggerate the spatial distinction between voice and environment in the name of realism. They construct sonic spaces differently from the character-driven docs and give the speaking subjects a different role in relation to the sonic environment. Ultimately, though, these two approaches are flip sides of the same coin. Both *Last Train Home* and *Roundabout in My Head* mix their sound to reconcile space with voice, and their soundscapes integrate the voice into environment to highlight the verisimilitude of each.

The spectral voice

The realist, integrative sound design of *Last Train Home* and *Roundabout in My Head* draws inspiration from fiction film sound design that carefully mixes voice, effects and ambient sound to balance verisimilitude and story information. Other documentaries, however, draw a different inspiration from the same fiction practice by taking more liberties with effects and mixing. These move away from the invocation of space to something more stylized. *Man on Wire* represents a practice that makes the human voice a haunting presence in its recounting of a historical event. Rather than a spatialized voice it constructs a spectral voice. Michel Chion (1999) theorizes the *acousmêtre*, the voice that is sonically present but visually absent, to have spectral qualities. Chion associates the *acousmêtre* with power and mystery, and while most fiction films have moments of *acousmêtre*, some exploit it more extensively. As René Throeau Bruckner summarizes in his reading of spectral sound in classical Hollywood, 'An *acousmêtre* amounts to a kind of phantom whose voice is not (yet) accompanied by a human body' (2015: 100).

Documentary cinema would seem to be removed from fiction films like *Testament of Dr. Mabuse* or *Psycho* that use the spectral, unseen voice. However, *Man on Wire*, a re-enactment-heavy documentary about tightrope walker Philippe Petit's 1974 walk between the World Trade Center buildings, exploits the expressive possibilities of the *acousmêtre* in its sound mix. It features present-day interviews with its subjects, Petit and his co-conspirators in the walk, as they describe the event and, in flashback, re-enactments of their preparation for it. However, the film is not a straightforward recollective documentary. Much like Errol Morris's *Thin Blue Line* (1988) or Andrew Jarecki's *Capturing the Friedmans* (2003), the film blurs fictional re-enactment and non-fiction elements in a stylized approach (Ricciardelli 2010). Like other films of this type, *Man on Wire*'s editing strategy is driven by suspense. Hamilton Carroll points out that the suspense is not straightforward: 'The audience of *Man on Wire* knows from the start how events turn out: Petit's wire walk is successful, he is arrested and then freed, he does not die in the attempt. The audience knows what; it is given a version of the how and the why' (2011: 839). Instead, the suspense lies in the 'how' of Petit's walking, as each stage poses a potential challenge to the mission of the walk; the spectator knows the eventual outcome but not the intermediate events.

The structuring absence in the documentary is the attack on and destruction of the World Trade Center buildings on 11 September 2001. The film makes no direct reference to the 9/11 attacks but the ultimate fate of the buildings acts as a haunting presence. As John Corner notes, 'As at other points in the film, the later tragic history of the Twin Towers contributes to the overall tonality, always staying in the background against depiction of the astonishing feat of 1974, but placing what we see against a known future and giving certain moments an added poignancy' (2014: 128). Corner argues that the scoring, with Michael Nyman's music, helps create the thematic opposition of time, a 'framework within which historical time is connected to durational time' (126). Similarly, Adam Ross Rosenthal (2012) sees the film's flashback structure as framing the historical account as a dream. Crucially, the film begins with a dream which Petit recounts, noting 'in the middle of the night, I have almost a nightmare'. He further connects nailing a crate of supplies to the nailing of a coffin. The scene seamlessly segues from dream to an account of the stunt preparation presented in re-enactment. The dream might be a different metaphor for cinema from the ghost, but here they work in tandem; by re-presenting the 1970s and the World Trade Center soon after their construction, *Man on Wire* implies a haunting of the present in the past, and the echo of Petit's stunt in the 2001 attack on the World Trade Center.

The film's sound design carries much of the weight of the film's sense of temporal haunting. Paradoxically, the quality of the voice recording is often too perfect for a documentary, as conventionally practised. Unlike the location and interview sound heard in *Last Train Home*, the interview sound in *Man on Wire* is often recorded in a sound studio or well-soundproofed room. It is 'cleaner' than documentary sound tends to be and has a rounded, sonorous quality that keeps it from seeming flat. As such, it is more akin to a studio or a radio broadcast recording. The clarity and sonority of the interview is enabled by the minimalist interview setting – rooms that minimize external noise yet provide a naturalistic, warm sense of echo. Clean sound free of ambient noise can, in the context of a fiction film, be compatible with a realist, unremarkable soundscape. However, in this case it separates Petit and his colleagues from the space of the footage, both archival and re-enactment, while environmental sound offers a greater sense of spatiality to the footage than is typical for historical documentaries. So, rather than tempering the *acousmêtre*, as is so often done, *Man on Wire* embraces the *acousmêtre*'s spectral qualities.

The opening scene demonstrates the complexity with which the film mixes voices and sounds in different registers of reality. The films starts with a black screen and a sonic fade in to sounds of raindrops, then thunder. Over the initial shot of a crate seen through a window, Petit recounts his dream in voice-over. His voice is sonorous, with a slight echo, but otherwise crisp in its recording; both the echo and the clarity emphasize the internal voice effect of the dream recounting. A sound montage then blends in another voice: that of Richard Nixon, giving his famous 'I am not a crook' speech during the Watergate crisis. Nixon's voice provides historical contextualization and also establishes the theme of the relationship between Petit's stunt and the political turmoil of the early 1970s. (Was his walk a form of escapism? Or a reflection of a zeitgeist?) However, Nixon's voice, readily recognizable in its deep timbre, is clearly different from Petit's in quality. It is a historical recording with the grain of archival sound – the television microphones of the 1970s and pre-Dolby magnetic tape effecting the voice's timbre and providing a slight archival hiss. *Man on Wire* exaggerates this archival grain by adding reverberation to the mix, forming a greater contrast to Petit's voice and to the sound effects or Foley of an unnamed colleague doing preparations.

Like surround-sound fiction films, *Man on Wire* emphasizes both sonic directionality and sonic perspective. In the example of the re-enactment described above, sounds correspond to what is shown. The televised speech dominates the sound mix in the left of the mix, whereas the sound of the conspirator placing an arrow in a concealing tube is located in the centre right; both sounds match the placement of their corresponding image in the frame. As the camera tilts down and tracks away, the timbre of the broadcast changes, becoming muffled and tinny to suggest its recession into the distance. Meanwhile, its place in the surround mix moves to the far left, while the colleague in the background moves to the far right. A woman in the foreground appears in the frame, writing on a padded envelope, with the sound of felt marker on paper particularly loud, nearly drowning out Nixon's speech. On the one hand, this sound design approach is less realistic, since it selectively emphasizes only very few sonic elements (people carry objects in the background but are inaudible, for instance). On the other hand, sound perspective lends a three-dimensionality that spectators often associate with verisimilitude, so *Man on Wire*'s use of stereo, while forced, is in some respects is more naturalistic than the radio-mic-heavy sound of either *Last Train Home* or *Roundabout in My Head*.

As in many fiction films, *Man on Wire*'s sound mix is highly exaggerated, with a subjective focus that favours some close sounds over a sonic portrait of the profilmic in its entirety. Mark Kerins has argued that, within the fiction film, digital surround sound has had the paradoxical effect of making movies louder but also softer. The noise reduction offered by Dolby and digital technology has increased the expressive possibilities of dynamic range in cinema, and Kerins notes, 'Digital surround's increased dynamic range allows increased flexibility in the use of "relative" silence, since even sounds played at a "normal" volume can be relatively silent in the context of a noticeably louder or busier soundtrack' (2010: 59). Similarly, while *Man on Wire*'s dynamics are not extreme its opening scene does rely on the sharp definition between loud and quiet parts of the dynamic range in a similar way to commercial fiction film. The mix, for instance, balances softer moments with the exaggerated 'close up' sounds of the television or the sound of arrow being placed in a tube.

The voices of *Man on Wire*'s subjects are spectral because they simultaneously invoke presence and absence. The overall sound design isolates the voice and gives each part of a multi-channel mix a distinct autonomy. The paradox of spectral sound is that a detailed sound mix aids the half-present, half-absent presentation of the voice. Petit and his co-conspirators speak with voices that are three-dimensional (they are recorded with room echo rather than with flattening radio mics) yet lack a clear spatiality. The acousmatic qualities of their voices suggest the power of the voice as a narrating agent and also its absence from a historical real that is represented. The sonic approach suits the visuals of re-enactment in *Man on Wire*, but the voice is not that of re-enactment per se. Rather, it stands outside the space of the image while making the image seem to emerge from the voice, as if the manifestation of a dream. The spectral voice invokes the real and suggests its haunting elusiveness.

The uncanny voice

Man on Wire's sound mix is much more stylized than documentaries with a spatialized voice, but it fundamentally does not break the illusion that sound emerges from the profilmic. Even when they add Foley and effects, the post-production additions buttress multiple sonic elements to create a verisimilar whole. In contrast, there are some documentaries that play with Foley to disrupt this illusion. Rather than placing voice in the cinematic space or suggesting a haunting quality, these films outright turn voice (and space) into an uncanny effect. In *Under the Sun*, for instance, Vitaly Mansky uses Foley sound effects to suggest the uncanny of the documentary real, particularly one highly controlled by the North Korean government. The North Korean government provided access to the filmmaker only on the condition that he shoot a fictional script of a 'typical' North Korean family as a 'documentary'. Faced with this restriction, Mansky decided to keep the fictional scripted material but to show the fake documentary as precisely a fake. *Under the Sun*, for instance, keeps footage of the stage handling by North Korean government workers, including their stage direction to the social actors. A dinner scene with the main character, a young girl named Zin-mi, and her family, for instance, begins with a shot of the family looking aimlessly around the room, waiting to begin their performance; the next shots include those of their handlers walking into the living room from offstage. These framing shots show the already artificial dialogue to be especially performative. This practice is in line with similar self-reflexive techniques in other documentaries that play up the artificial quality of documentary filming – for instance, in the way Ross McElwee re-stages a filmmaker–subject encounter twice in *Six O'Clock News* (1996).

Under the Sun uses sound design to compound the reflexive commentary on the underlying propaganda material. Drawing inspiration from recent experimental documentary makers, such as Marie Voignier (whose *Tourisme Internationale* [2014] also uses Foley to question North Korean propaganda), Sergei Loznitsa (*Blockade* [2006]) and Andrej Ujica (*The Autobiography of Nicolae Ceausescu* [2010]), Mansky and his sound designer Anrijs Krenbergs add a considerable amount of Foley sound to otherwise mute images. While Foley in other documentaries may be part of a composite realist soundtrack, its use from the beginning of *Under the Sun* is unusual. As Zin-mi boards a bus in an opening sequence, the film presents no sound except for Foley of cars and cyclists passing, the soundtrack otherwise eerily quiet. At other points, the film presents crowds with only walla and overly perfect Foley of rushed footsteps. The mismatch of sound and image challenges the voice of the North Korean versions of events, literally and figuratively, while giving the represented spaces of Pyongyang an unreal quality.

The Foley is in many respects too perfect a match for the image. Footstep sounds only appear when the image shows people walking or a bus's appearance cues its presence on the soundtrack. Rather than create a superfield, the sound here is uncanny in its neatness. The spectator may realize the sonic mismatch or misapprehend it entirely. Like a visual cousin of the uncanny valley (in which partially verisimilar representations seem real but not quite), the uncanny sound is perceptually challenging because it is almost the expected sound yet not quite.

The film juxtaposes three different types of voice recording. Voices from public address loudspeakers echo through the film's exterior spaces, as a reminder not only of the state's official voice but the lack of access to any unplanned subject interaction or testimony. The walla too functions as an uncanny voice, a kind of vox populi more powerful for its ultimate indeterminacy. In contrast, Zin-Mi and other of the main social actors in interior scenes have clear, resonant voices, since the controlled filming conditions have allowed for minimum sonic interference. However, rather than serve as the authentic alternative to the anonymous voice of the state or the crowd, the overly clear sound oddly fictional by being too perfect a capture of the vocal performance.

Under the Sun is a film that attaches a thematic importance to voice. The film ends on a tightly framed shot of Zin-mi as she answers a teacher's question about joining the Children's Union, a para-state youth organization. She starts to recite the answer in a clear, loud declarative voice but stumbles over a forgotten line. The affect in her voice and face, tortured and mortified, stands out as the unplanned reaction that Mansky was not able to record during the rest of his filming (or decided not to show). Prevented from capturing the social actor in a manner that he might expect to in another national context, Mansky embraces the limitations of the recording by questioning its authenticity on the soundtrack. The effect is all the more striking as the teacher asks Zin-mi to recite a cheerful poem. Recovering, the girl launches into a patriotic poem with a flat vocal affect, communicating its rote quality. Rather than confront the image directly with the soundtrack, *Under the Sun* deconstructs it. As the spectator frequently encounters only a fictive echo of the sound the visuals invoke she therefore is encouraged to question the image. The intrusion of the 'real' voice is powerful because it can never take the authentic place of the uncanny voice from the rest of the film. The uncanny voice challenges the expectations of verisimilar sound by reducing the spoken word and environment to mere illustrations of the visual, thereby undermining the signifying stability of both sound and image.

Conclusion: Silences of the voice

Sound has held an important role in documentary theory, in part because of the historic importance of voice-over narration and social actor testimony for anchoring the meaning of the image. This relation of literal to figurative voice has informed theory from Pascal Bonitzer (1986) to Bill Nichols (1993). However, the broader analysis of the sound–image relationship has not always opened the way for a granular account of sound design as a creative practice. In response to the historic neglect of the soundtrack, sound studies as a subfield has argued for the important, even determining, role of film sound in the overall meaning and aesthetics of cinema, and recent historical work on fiction cinema (particularly Hollywood) has further chartered the interaction between industrial practice, film sound technology and sound aesthetics. Bringing these areas of inquiry together – documentary theory and the history of sound design – can help historicize and illuminate contemporary documentary practices. The human voice in documentary is such a rich part of the documentary soundtrack because it serves in dialectical relation to the environmental sounds that invoke space in non-fiction cinema.

This chapter has argued that in the age of digital sound, the voice has tended towards three orientations regarding its sonic environment: the spatialized, the spectral and the uncanny voice. These categories of sonic construction can help explain how individual documentaries make use of the human voice. Moreover, they are a reminder that the 'voice' in documentary is not simply the result of the recording of the human voice but also how an overall sound design constructs the voice in relation to the environment. Whether a documentary focuses more on its characters or the natural and manmade environments its subjects inhabit, the sound design situates the space and verisimilitude of voice for the spectator. Sound designers have an unlimited number of choices but in practice their design tends towards recognizable narrational forms and styles that are inspired in part by fiction cinema.

The distinct variations of the spatialized, spectral and uncanny voice suggest the range of documentarians' response to newer environmentally oriented sound design practices. Each aesthetic construction of the voice has emerged in response to digital sound, which has inherited some of the challenges of analogue documentary sound (with the focus on achieving clarity) and added the expressive possibilities of easier and more affordable post-production capability. The latter has not made a convergence with fictional sound practice inevitable but it has certainly encouraged that merger. The result is a paradoxical situation for contemporary documentary aesthetics: the treatment of the voice in recent documentaries draws inspiration from fiction cinema, while the voice's role in a film's overall meaning retains a distinct role specific to documentary.

Note

1 For the first wave of sound studies scholarship, see Weis and Belton (1985) and Altman (1992). Michel Chion's work (1999, 2009) is, as often noted, foundational for theorizing film sound. More recent film histories on Hollywood sound include Kerins (2010), Beck (2016) and Kalinak (2015).

References

Altman, R., ed. (1992), *Sound Theory, Sound Practice*, New York: Routledge.

Beck, J. (2016), *Designing Sound: Audiovisual Aesthetics in 1970s American Cinema*, New Brunswick, NJ: Rutgers University Press.

Bonitzer, P. (1986), 'The Silences of the Voice', in P. Rosen trans., eds M. Butzel and P. Rosen, *Narrative, Apparatus, Ideology*, New York: Columbia University Press.

Carroll, H. (2011), 'September 11 as Heist', *Journal of American Studies*, 45 (4): 839.

Chion, M. (1994), *Audio-Vision*, New York: Columbia University Press.

Chion, M. (1999), *The Voice in Cinema*, New York: Columbia University Press.

Chion, M. (2009), *Film, A Sound Art*, New York: Columbia University Press.

Corner, J. (2014), 'Music and the Aesthetics of the Recorded World', in H. Rogers (ed.), *Music and Sound in Nonfiction Film*, 123–136, New York: Routledge.

Ferhani, H. (2016), Question and Answer Interview, Lincoln Center, New York, April 15.

Flanagan, M. (2008), 'Towards an Aesthetic of Slow in Contemporary Cinema', 16: 8. http://www.16-9.dk/2008-11/side11_inenglish.htm.

Heath, S. (1982), *Questions of Cinema*, Bloomington, IN: Indiana University Press.

Kalinak, K., ed. (2015), *Sound: Dialogue, Music, and Effects*, New Brunswick, NJ: Rutgers University Press.

Kerins, K. (2010), *Beyond Dolby (Stereo): Cinema in the Digital Sound Age*, Bloomington, IN: Indiana University Press.

Leacock, R. (1961), 'For an Uncontrolled Cinema', *Film Culture*, 22–23: 23–25.

Leonard, M. and R. Strachan (2014), 'More than Background: Ambience and Sound-Design in Contemporary Art Documentary Film', in Rogers, H. (ed.) *Music and Sound in Nonfiction Film*, 166–179, New York: Routledge.

Negt, O. and A. Kluge (1993), *Public Sphere and Experience: Toward an Analysis of the Bourgeois and Proletarian Public Sphere*, trans. P. Labanyi, J. O. Daniel, and A. Oksilaff, foreword by M. B. Hansen, Minneapolis, MN: University of Minnesota Press.

Ricciardelli, L. (2010), 'Documentary Filmmaking in the Postmodern Age: Errol Morris & The Fog of Truth', *Studies in Documentary Film*, 4 (1): 35–50.

Rosenthal Adam, R. (2012), 'Suspended Reading: *Man on Wire*, 9/11, and the Logic of the High-Wire', *Screening the Past* 8, web.

Ruoff, J. (1992), 'Conventions of Sound in Documentary', in R. Altman (ed.), *Sound Theory, Sound Practice*, 221, New York: Routledge.

Throeau Bruckner, R. (2015), 'Bad Sync: Spectral Sound and Retro Effects in Portrait of Jennie', in M. Leeder (ed.), *Cinematic Ghosts: Haunting and Spectrality from Silent Cinema to the Digital Era*, 97–114, New York: Bloomsbury Academic.

Weis, E. and J. Belton, eds (1985), *Film Sound: Theory and Practice*, New York: Columbia University Press.

12

Kazuhiro Soda's Migrant Voices

Luke Robinson

Introduction

This chapter addresses the problem of how to 'give voice' to the documentary subject. Traditionally associated with ideas of agency and subjectivity, and central to both the observational and participatory filmmaking traditions, contemporary criticism emphasizes that 'giving voice' can instead lead to forms of objectification. This is primarily because the practice is not neutral, but shaped by asymmetrical power relationships between documentary-maker and subject. Such a tension presents particular problems for filmmakers seeking to document socially marginal people. While these projects are often explicitly driven by a desire to magnify the voices of the voiceless, their director–subject relationships are among the most unequal. For many filmmakers, the advantages of the former outweigh the dangers of the latter and these directors often continue to shoot their subjects using comparatively traditional observational and participatory filmmaking techniques, which convey particular formal and institutional advantages. But other filmmakers have sought to tackle this problem by deconstructing or challenging the very idea of the voice, either through abandoning recognizable documentary modes in favour of experimental filmmaking or adapting and experimenting within the limitations of the documentary tradition.

Here, I explore how the Japanese director, Kazuhiro Soda, addresses this conundrum in his film, *Oyster Factory* (2015). Rather than directly 'giving voice' to the Chinese migrant workers at the heart of the documentary, *Oyster Factory* instead combines strategic subtitling with sound techniques standard to observational cinema to suggest how 'having' or 'giving voice' is itself a socially constructed process privileging certain groups over others. At the same time, by interweaving and juxtaposing different voices across the film, Soda suggests ways in which the viewer might try to understand migrant experience, without ever speaking for the documentary's subjects themselves. Borrowing from theorist Trinh T. Minh-ha, I term this practice a form of 'speaking nearby': a way of 'giving voice' that does not speak for, but to, those with whom it engages.

The problem of 'giving voice'

The idea of 'giving' or 'having' a voice has a long history in documentary filmmaking. Traditionally, it has been intimately connected to beliefs regarding documentary authenticity and efficacy. With the advent of sound technology, the filmed interview and then recorded everyday speech became markers of immediacy and documentary authenticity. What Brian Winston (2008: 51) calls 'letting people do their own talking' quickly became 'a crucial development for subsequent realist documentary work'. At the same time, Laura Kunreuther (2014: 36) stresses that 'the voice in modernity typically works [as] an index of interiority, consciousness, agency, and self'. Consequently, the act of speaking is often positioned by Western filmmakers and critics as 'a liberatory act of giving expression to an interior idea, thought, opinion, or wish that inaugurates the subject's entrance into the political sphere and, indeed, into humanity' (Rangan 2015: 100). Michael Renov (2004: 201), for example, notes the contemporary emergence of the immersive, first-person video confession 'that obsessively tracks personal truths', while also tracing the roots of this form partly back to Jean Rouch and Edgar Morin's experiments with interview, soliloquy and encounter in *Chronicle of a Summer* (1961) (197). Here, the voice in moving image production is understood as a marker both of authenticity – although of a different kind from that associated with the interview – and of a specific type of subjectivity, one that is both political and individual.

Of course, these assumptions are problematic. Produced as it is by the body, the voice is mediated before it even reaches the camera. Classic techniques through which voice is presented in non-fiction, such as the talking head, are therefore often forms of performance, even if they have not always been recognized as such.[1] The kind of subjectivity identified with the voice is itself quite particular, as implied above. Usually, a self-conscious, problem-solving individual is assumed, one whose hidden depths are revealed through the act of speaking: the 'ideal person of neo-liberal ideology' (Kunreuther 2014: 30). This issue is not necessarily overcome by simply turning the camera over to one's subjects, for participatory production itself can be predicated on the existence of a similar subject, or work towards its production. The 'empowerment' of minority subjects, encouraged to document their own lives on camera, can easily subsume political agency under entrepreneurship, especially when the images produced are then sold on the open market.[2]

Finally, the belief that the voice is the route to subjectivity ignores the ways in which it may be a mark of objectification. Pooja Rangan (2015: 98–99), drawing on Rey Chow's (2014) reading of Frantz Fanon (1967), points out that for some, the voice may subjugate as much as emancipate. Not only does the spoken word have the capacity to objectify, as with racial insults, but traces of embodied difference embedded in the voice – accent, for example – may also be used to dehumanize certain speakers in certain contexts (Rangan 2015: 103). Clearly, some people can both give and have voice more easily, and on their own terms, than others. The conditions of possibility under which the documentary subject gets to speak, or be heard, are therefore determined both by technologies of mediation and by social dynamics.

This tension between subjectification and objectification that lies at the heart of 'giving voice' presents particular problems for directors working with people at the social margins. Frequently, such filmmakers work with minority groups precisely to amplify their voices. But under the conditions outlined above, achieving this amplification without speaking for such groups, or forcing them to speak in ways that exacerbate this marginalization, is necessarily a fraught task. One response to this conundrum has been the adoption of techniques that reflexively problematize the very idea of the voice. Rangan addresses the work of Leslie Thornton, whose experiments with wordless or multiply voiced soundtracks expose the limits of the voice-over as a political tool within the feminist documentary tradition (2015: 108–123). Another approach is the animated interview. As Annabelle Honess Roe (2013: 88–104) has noted, this technique both protects the interviewee from visual identification – critical if this subject is politically or legally vulnerable – and directs us to reflect more closely on the non-verbal elements of the voice that are both critical to its constitution and often overlooked when the interviewee is visible on screen. In both of these examples, the desire to 'give voice' is balanced out, even superseded by, a desire to explore how this practice is formally mediated and socially constituted, in the process highlighting the conditions of possibility under which certain groups get to speak, or be heard, at all.

Filmmakers using the techniques discussed above achieve reflexivity by explicitly decoupling sound and image, either literally (the voice paired with animated rather than photographic imagery) or figuratively (a voice or voices whose relationship to the image is hard to interpret). This disruptive practice runs counter to the tenets of observational filmmaking. 'Letting people do their own talking' has been central to the latter mode ever since its emergence in the 1960s, and, despite systematic criticism of its many shortcomings, observationalism remains a stubbornly persistent documentary practice. The practice's advantages – among others, its association with documentary 'authenticity'; a tradition of institutional and social analysis; the comparative ease with which it allows for longitudinal study of one's subjects – perhaps explain why many filmmakers still adopt it to film 'at the margins', whether these be the impoverished rural China of Wang Bing's *Three Sisters* (2012), or the island of Lampedusa, at the vanguard of the European migrant crisis, in Gianfranco Rosi's *Fire at Sea* (2016).

This does not mean that observational cinema has remained static: films such as Laura Poitras's *Citizenfour* (2014), or Lucien Castaing-Taylor and Véréna Paravel's *Leviathan* (2012), rework the tradition for a contemporary moment. But *Citizenfour* reflexively interrogates the visual relationship between observation and surveillance, while Castaing-Taylor and Paravel explore an immersive aesthetic that seeks not to decouple sound and image but rather connect sight and sound to the other senses. Neither film deals with socially marginal subjects as such, which perhaps explains why they do not explicitly address the voice as a documentary trope.

Is it possible, then, to defamiliarize the act of 'giving voice' to the socially marginal documentary subject such that its social and technological construction becomes obvious, without necessarily departing entirely from the conventions of observational filmmaking? I believe it is, and the film I want to use here to illustrate this possibility is *Oyster Factory* (2015), by Japanese director Kazuhiro Soda. At the heart of *Oyster Factory* is a cross-cultural encounter between majority and minority groups: oyster farmers in the Japanese countryside and their mainland Chinese seasonal labourers. Japan and the PRC have a fraught relationship, overshadowed by Japan's invasion of China in the 1930s, its refusal to publicly apologise for war crimes since and the continued exploitation of this history by politicians in both countries for domestic gain. Although I argue that this shared history overdetermines the position of these Chinese workers vis-à-vis their Japanese employers and co-workers – a colonizer–colonized dynamic that is built on a historical 'othering' of the Chinese by the Japanese, despite (or perhaps because of) their many shared cultural practices – Soda uses a combination of observational sound practice and tactical subtitling to implicate the film's audience in this process of othering, suggesting how voice is socially constructed – and restricted. Finally, I explore how the director's framing of a key Japanese figure in the film, Watanabe, recuperates the objectification of these Chinese migrants through a process similar to Trinh T. Minh-ha's concept of 'speaking nearby'. The result is a film that avoids speaking for, or over, its migrant subjects, while still operating within the boundaries of a recognizably observational style.

Oyster Factory: '"China" is coming'

Oyster Factory is set in a ten-person seafood processing plant in rural southwest Japan. The Hirano Oyster Factory is located in Ushimado, a little town perched on the edge of the Seto Inland Sea, the body of water that connects three of Japan's four primary islands. Owned by Hirano, after whom it is named, the factory is now to be run by his successor, Watanabe. Consequently, it is no longer a family business; Hirano's son, who makes a brief cameo appearance late in the documentary, has rejected the life of an oyster fisherman in favour of the more stable, less physical, career of the classic Japanese salaryman.

This transition in ownership is a small detail that illuminates the broader theme of the entire documentary: migration. As younger Japanese like Hirano's son leave the countryside in search of life and employment in larger cities, so rural towns like Ushimado suffer systematic depopulation and resultant labour shortages. In turn, small businesses such as the Hirano Oyster Factory, and other seafood plants in the area, turn to migrant labour from overseas to fill the gaps. This is the point of departure for Soda's documentary. Ushimado is the hometown of Soda's wife and producer, Kiyoko Kashiwagi (Tsui 2016: n.p.), and a place the couple holiday regularly (Elphick 2016: n.p.). Initially drawn to the fisherman precisely because they worked in a dying industry, Soda started filming at the oyster factory, only to discover that, for the first time ever, they were going to take on two temporary workers from China, via a Japanese employment agency. As he puts it, 'So all of a sudden, it [the film] became a story about globalization. A small oyster factory became a microcosm of globalization' (Curran Bernard 2015: 327–328).

In consequence, *Oyster Factory* is structured around preparations for the arrival of the Chinese workers, followed by their acclimatization to their new surroundings. At the beginning, we are introduced to Watanabe at work on his boat, bringing in nets of shellfish. Having established the labour of fishing, the camera then follows him back to the plant. Here, we get the first intimation of the documentary's main focus: a calendar on the factory wall is marked with 'Saturday, 9th. "China" is coming'.[3] From this point on, Soda shadows the everyday patterns of both local fishermen and factory workers; explores stories about the existing community of migrant workers in Ushimado; digresses through a narrative strand about a cat who constantly visits the director and his wife for food; and tracks Watanabe as he prepares to receive his labourers, going so far as to buy them a prefabricated unit as housing. When the two men arrive, we witness Watanabe show them around the factory; introduce them by their names (Zheng and Zhao) to his Japanese employees and to other Chinese workers from neighbouring plants; demonstrate how to cook and eat oysters; and, finally, introduce them to working on the boat. The film concludes here, on their first day of work, as it becomes abundantly clear neither man has ever been to sea before.

Soundscapes of Japanese racism

As with the director's previous films – *Campaign* (2007) and *Campaign 2* (2013), *Mental* (2008), *Peace* (2010), *Theatre 1* and *Theatre 2* (both 2012) – *Oyster Factory* is primarily, though not strictly, observational in style. Sound is therefore naturalistic and location recorded; voice-over is eschewed entirely. Although Soda claims to be heavily influenced by Frederick Wiseman (Curran Bernard 2015: 318), he does not entirely eliminate himself from the diegesis: at times the director interacts with his subjects from behind the camera, asking questions and making conversation. In one early, reflexive encounter at a seafood market, an old woman looks directly at camera, points at produce and asks, 'How much? How much is this?' before saying, 'Oops, it was a cameraman' and scuttling away in embarrassment, with Soda laughing loudly from off screen. The voices of his subjects thus emerge through everyday conversation and the filmmaking encounter, not via formal interviews.

Much of what we learn from Soda's Japanese subjects helps paint a picture of everyday life in Ushimado, whether this is through discussion of the fishermen's working day, or of Shiro, the white cat who has 'adopted' Soda and his wife, and whose owner apparently has four cats like him. But in the first two-thirds of the documentary, what also emerges through such digressions and asides is how these Japanese view the migrant Chinese labourers working in the local fishing industry. These attitudes are not uniform; indeed, the comments Soda captures suggest they are highly gendered. One of the Japanese women factory workers, for example, has nothing but praise for her Chinese (women) colleagues, whom she notes are hard workers who pick up shellfish shucking with ease. But feelings among the male factory managers are overwhelmingly negative. Our first exposure to these views comes thirty minutes into film, when Watanabe is talking to another factory manager. The latter is complaining about a Chinese worker who has quit after only a day; the factory owner characterizes this man variously as 'not capable', 'lazy' and 'mental', before going on to say that he'll just have to work the remaining two workers twice as hard, to make up the difference. A little later, an electrician compares the Chinese migrants to the Vietnamese workers whom they replaced. While the Vietnamese were 'nice people' who simply couldn't take the Japanese weather, the Chinese, he says, are 'terrible … Simply put … they just steal whatever they see … They don't care … That's why when the Chinese stay in a hotel, hotel managers empty the room as much as possible. They take away all the expensive furniture to prevent them from stealing'. Finally, on the day the migrant labourers are due to arrive at the Hirano Factory, Hirano himself suggests to Soda that it would be better if he stopped filming, in case he enrages the Chinese: 'There was even a homicide in Hiroshima. It would be trouble if they don't like your shooting.' In these exchanges, the explicitly racist stereotypes that many locals hold of these outsiders are made very clear indeed.

If we position *Oyster Factory* as a study of globalization's impact at a local level, it would seem most obvious to frame this racial panic as a by-product of concerns about job security and Japanese deindustrialization. But the distinction drawn above between Vietnamese and Chinese workers hints at a longer, more historically overdetermined dynamic. Although the Japanese Imperial Army occupied French Indochina for five years during the Second World War, the Japanese occupation of China lasted almost three times as long, and was considerably more bloody.[4] The discourse legitimizing this occupation positioned the Chinese as culturally deficient and racially inferior (Stegewerns 2003: 109), resuscitating old tropes of Chinese cowardice and asserting newly racialized ones stressing their innate venality and dishonesty (Young 1999: 163–167). The refusal of successive Japanese governments to address the country's imperial war crimes, combined with the Chinese Communist Party's exploitation of this grievance for a variety of political ends, means that divergent memories of this period are ritually invoked by nationalist groups on both sides.[5] In these circumstances, it is hard not to hear, in criticisms of Chinese workers as lazy, mentally unhinged or inherently untrustworthy, not merely worries about impending unemployment of the decline of the Japanese fishing industry, but also the echoes of a longer, far more brutal history that continues to frame interactions between citizens of both countries. Soda's filmmaking captures the sights and sounds of quotidian life in present-day Ushimado, but in the process it also reveals how darker dynamics from the past continue to shape the everyday present.

'Giving voice' to the Chinese migrant

If Soda's broad adherence to (modified) observational practices allows him to draw out more than might be expected from his Japanese compatriots, it shapes what we, as viewers, learn about Hirano Oyster factory's Chinese migrants in rather different ways. Soda speaks no Chinese; the two men speak the most basic Japanese. Upon the latter's arrival, it quickly becomes apparent that the director can only directly engage them in very limited conversation. At his first and sole attempt, he asks, in English, 'Do you speak English?' After Kiyoko Kashiwagi, positioned behind the camera, translates the word 'English' into Japanese, one of the men responds, 'En-gu-li-sh? Yes! No!' All concerned then burst out laughing. Linguistic difference limits what Watanabe, the factory manager, can communicate to his new employees as well as circumscribing how the latter interact with their Japanese co-workers. In such circumstances, all involved rely primarily on body language and simple instructions. In effect, the language barrier, combined with the exclusive use of location sound, minimizes both the migrants' own voices and any indirect conversation through which we might learn about them. Yet Soda pushes these limitations to extremes in other ways. While the labourers cannot converse with the filmmaker or with the locals, they can, and do, converse with other Chinese migrants. As Zheng and Zhao's first day draws to a close, they gather with three other Chinese labourers from nearby factories and, standing around Watanabe's truck, chat at length, with the manager's children playing games around them. What the migrants are discussing, however, we do not know: Soda does not subtitle their dialogue.[6] Indeed, for the remainder of the documentary, all conversations in Chinese remain unsubtitled, as inaccessible to a Japanese audience as to an Anglophone one.

This move would be striking in any documentary, but it is particularly unusual given that, narratively speaking, *Oyster Factory* builds up to Zheng and Zhao's arrival. One thus expects to find out more about them directly, to hear from them in their own words. Instead, our epistephilia is explicitly denied; what little we learn about them is conveyed to us through their interpreter, who is present only on the afternoon the workers arrive. We can understand this decision as straightforwardly realist in rationale. According to Soda (2017: n.p.), 'I made a choice not to subtitle them because that's how I experienced these scenes while I rolled the camera'. Refusing to provide the audience with privileged access to the film's subjects can thus be interpreted as reinforcing the immediacy of the viewing experience. But I would argue that it is also a reflexive technique, one that denaturalizes the idea of giving voice. In her essay 'Mechanical Eye, Electronic Ear, and the Lure of Authenticity', Trinh T. Minh-ha points out that subtitling the foreign subject in sync sound 'factual films' is *de rigueur*:

> IT [their voice] HAS TO MAKE SENSE. WE WANT TO KNOW WHAT THEY THINK AND HOW THEY FEEL [capitals in original]. Making a film on/about the 'others' consists of allowing them paternalistically 'to speak for themselves' and, since this proves insufficient in most cases, of completing their speech with the insertion of a commentary that will objectively describe/interpret the images according to a scientific-humanistic rationale. (Trinh [1984] 2016: 760–761)

In other words, subtitling makes sense of the 'other' by translating his or her voice into a language that we can hear and understand, in the process erasing the difference between audience and filmed subject. We could say that it functions to interpellate the speaker's subjectivity in terms that we comprehend. In contrast, Soda's Chinese subjects conserve their original voice, and with it that critical element of difference. They speak, but we cannot understand.

This preservation of difference has a number of consequences. In Trinh's terms, the Chinese migrant labourers retain an agency independent of both their Japanese employers and the extra-diegetic audience. In Ushimado, they create a space between themselves through language which cannot entirely be managed, a space that is then recreated in the documentary's diegesis. It also allows us to turn our attention to the non-linguistic elements of conversation – to what Rangan (2015) terms the 'matter of the voice' – to consider the importance of tone and timbre to the meaning of voice, since the words themselves are not translated. But, most importantly, I think this moment throws into relief the film's exploration of the social construction of voice *outside* the imposition of direct speech, and the epistemic violence, the 'othering', involved in this process, while simultaneously positioning the audience as active participants in said othering. By the time Zheng and Zhao arrive, our expectations for their reception have already been shaped by the racialized commentary that has emerged from Soda's interactions with the Japanese factory managers in particular. In capturing these sentiments as they surface, *Oyster Factory* hints at just how hard it is for Chinese workers in Japan to speak or be heard outside long-standing, racialized frames of reference. If speaking – 'having voice' – presupposes humanity, then to speak as Chinese under these conditions is near impossible, for 'Chinese-ness' is itself understood as a mark of inferiority, both cultural and racial.[7] At the same time, the transition to unsubtitled Chinese-language dialogue towards the end of the film positions the audience as active participants in this process. By aligning the viewer's perspective with that of its Japanese subjects at a point when the audience is already highly aware of the existing discourse around these seasonal migrants, *Oyster Factory* implicates us in the dynamics of racial objectification. As Soda notes (2017: n.p.): 'Without subtitles, [a] non-Chinese audience would be put into a similar psychological state with the Japanese factory workers. Because viewers don't know what the Chinese are talking about, they start to speculate what they are saying – often negative imaginations, which creates uneasiness'. For example, at the end of the first day, two of the Chinese migrant labourers break away from the group and, at some distance from the camera, start to talk in lowered voices. Clearly, they are seeking some privacy. Yet, given all that we have heard to date, they also appear to be positioned ambiguously: perhaps they actually have something to hide? No judgement is made, and we simply observe the two men in conversation, but the space for this kind of projection is created. At moments like this, as the camera lingers on the two men talking quietly in the distance, Soda seems to be both replicating the dynamics that lead to vocal erasure and encouraging us to reflect upon our own contribution to this process.

Speaking nearby

Oyster Factory explores how voice is socially constructed, and how both its Japanese subjects and we, as viewers, might participate in the processes that limit who can speak and how they speak in public. We can understand Soda's particular combinations of sound and subtitling as a reflexive practice, one that resists the objectification of the Chinese workers through 'giving voice', while also encouraging us to think critically about the limits of this practice for migrant labourers in Japan. However, even if the documentary allows for a meditation on the dynamics of social inclusion and exclusion, there is still the danger that, by the end of the film, its Chinese subjects remain for us mere ciphers – even if we are aware of how and why this has come about. Can the film encourage us to view them differently, without 'completing their speech' in the manner that Trinh T. Minh-ha associates with inevitable objectification? I would suggest it can and that the key to this is the figure of Watanabe, the Hirano Oyster Factory manager.

While all the other Japanese in *Oyster Factory*, bar Soda and his wife, are clearly from Ushimado or the surrounding area, Watanabe is not. This is made apparent early in the documentary, in a sequence in which Soda is riding in a car with Hirano, the original owner of the eponymous seafood plant. Watanabe, Hirano explains, is actually from Miyagi Prefecture, in Japan's northeast. After a short pause, Soda asks 'Is he a nuclear refugee?' Hirano responds, 'Yeah. Actually it's not nuclear, it's because of the tsunami. He moved here to work in Mushiage [a nearby town]. But because I'm retiring, I asked him to take it [the factory] over'. As Watanabe himself makes clear to Soda in a later sequence, his hometown, Minamisanriku, was almost entirely destroyed by the 2011 Tohoku earthquake and tsunami. Displaced, he moved his family first to Tokyo, and then down south, where his skill as an oyster fisherman could be put to better use. This is where he met Hirano. Watanabe may be Japanese, but he is both an internal refugee and an economic migrant.

In consequence, over the course of the film, Watanabe provides an indirect point of entry into the exploration of the migrant experience without speaking for, or over, his Chinese employees. Unlike more minor characters who suggest the possible pluralization of the discourse around migration, Watanabe appears throughout the documentary. Through his conversations with other managers, with Soda and his asides to the camera, we build up a sense of his personal history and his emotional world. It is this indirect commentary that complicates the distrust and contempt with which the other factory owners all too clearly view the migrant workers. When, early in the film, he is engaged in conversation about the Chinese labourer who has quit his job early and unexpectedly, Watanabe's immediate response to the news is, perhaps somewhat surprisingly, 'Homesick?' When Soda asks why it is so hard to bring Japanese labour to Ushimado, he replies, 'People who would come here are kind of losers, aren't they?', an answer that suggests a complex relationship with his own move from Miyagi to the southwest. When questioned as to whether the temporary accommodation he builds for his workers is similar to that which he lived in after the tsunami, he explains it is better, because it is thermally insulated. And, while Hirano himself seems quite happy to talk about the overseas labourers as 'China', literally dehumanizing and objectifying them, Watanabe is the one who introduces them by name. It is gestures like these that, over the course of the documentary, gradually set him apart from his compatriots.

Watanabe's words and actions open a window onto his own migrant experience. He invokes the affective consequences of his own movement – homesickness, self-doubt, alienation – as well as its material hardships. At the same time, his experience speaks to the emotional world of his Chinese labourers; the very fact that it often emerges during discussions over the local Chinese migrants, or in preparation for Zheng and Zhao's arrival, makes this clear. And yet, Watanabe never speaks *for* these workers. Neither he, nor Soda, attempts explicit comparisons between his background and theirs. At the end of the day, Watanabe remains their manager, a professional relationship that is necessarily unequal; a more direct assessment would obscure these differences. Instead, his remarks illuminate their world at one remove, analogously, without pretending to be its mirror, or to bridge the many divisions that separate them, from language to social status.

This juxtaposition of experiences across the course of the documentary, the implicit dialogue between them and the work the audience has to do to make these connections sits well with Soda's own description of his filmmaking. He cites the root of the Japanese for 'observation' – *kansatsu*, literally 'looking' and 'sensing' – to argue that the observational style is not per se distancing, but rather encourages sustained engagement on the part of the viewer (Curran Bernard 2015: 314). Following Bill Nichols, we might understand the perspective that emerges from this formal layering as an example of the documentary voice or the text's 'social point of view' (Nichols 1983: 18); in the case of *Oyster Factory*, to borrow Trish Fitzsimmon's (2009) reformulation of Nichols's terminology, this voice could be described as dialogic or even choric, rather than monological.

But I would like to suggest another way in which we could frame the structure of *Oyster Factory*, one that also acknowledges the importance of voice in the more literal sense: Trinh T. Minh-ha's 'speaking nearby'. Trinh describes this practice as 'a speaking that does not objectify, does not point to an object as if it is distant from the speaking subject or absent from the speaking place. A speaking that reflects on itself and can come very close to a subject without, however, seizing or claiming it'. (Chen 1992: 87)

The paradigm for 'speaking nearby' is usually understood as the self-conscious, critical voice-over, such as that used by Trinh in *Reassemblage* (1982), her critique of anthropological filmmaking traditions (Mayne [1990] 2002: 101). Voice-over or no, I would argue that it is an appropriate description of how Soda positions Watanabe vis-à-vis his Chinese employees in *Oyster Factory*. Watanabe's commentary is used to reflect indirectly on potential parallels between his own life and that of his employees. Without substituting for the voices of these migrants, or rendering them transparently comprehensible, his own dialogue communicates a degree of proximity to Zheng and Zhao, yet never 'claims' kinship with them. Trinh herself largely dismisses observational filmmaking as objectifying as much as objective; this is one thesis of 'Mechanical Eye, Electronic Ear, and the Lure of Authenticity'. Here, though, despite his broad adherence to this form, Soda finds a way to respect the limits of 'giving voice' without entirely erasing any trace of the migrant experience from his documentary.

Conclusion

My aim in this chapter has been to explore some of the problems associated with the idea of 'giving voice' in documentary film. Tied up with issues of subjectivity and agency, and with objectification and exclusion, the question of how, if at all, one can provide one's subjects with a voice is a thorny one. I have argued that *Oyster Factory* is an example of a documentary that addresses this issue in several ways. First, by refusing to subtitle the film's Chinese dialogue, Soda allows his subjects to retain a degree of 'voice' and agency. At the same time, by layering this over a background of suspicion and xenophobia, he shows how this aporia could be interpreted through long-standing racialized categories, effectively interpellating the Chinese migrant in Japan before he or she has even begun to speak. Furthermore, he deliberately implicates the audience in this process. Finally, through the figure of Watanabe, the film develops what I argue is a version of 'speaking nearby': the articulation of a parallel migrant experience that does not stand in for that of its Chinese subjects, but nevertheless speaks to it from close by. The documentary therefore tries to explore the experience of its minority subjects without either speaking for them, or forcing them to speak our language – literally and figuratively – but also without entirely abandoning the idea that the migrant has a voice. Instead, it uses observational filmmaking to suggest the social dynamics that work to silence this voice, while also exploring alternative ways in which we can learn to listen for its resonances.

Notes

1 For further exploration of the talking head as a form of performance, see Martineau (1984) and Sarkar and Walker (2010).

2 See, for example, Pooja Rangan's (2011: 150–155) discussion of this dynamic in relation to participatory image production with children, and her exploration of how it informs the project 'Kids with Cameras', and Zana Briski and Ross Kaufman's documentary *Born into Brothels* (2004).

3 'China' here is a literal translation of the Japanese *kanji*, which are those for the country.

4 Although the Second Sino-Japanese War did not officially start till the 1937, the PRC government marks the occupation from the July 1931 Marco Polo Bridge Incident, which signalled the Japanese invasion of Manchuria, in northeast China. The First Sino-Japanese War occurred in 1894–95 and resulted in Japan's annexation of Taiwan, then officially part of the Qing Empire.

5 This happens most obviously during protests by right-wing political groups in Japan and during state-sanctioned anti-Japanese demonstrations in the PRC.

6 This is true of both the original Japanese and English-subtitled releases of the documentary. The former does not subtitle this discussion in Japanese; the latter, though it subtitles all Japanese conversation into English, leaves these Chinese-language exchanges untouched.

7 This underlines Rangan's contention that ethnic minority speakers are often placed in a double bind: to have a voice, they must lose 'the embodied position they allegedly speak for' (2015: 103).

References

Chen. N. (1992), '"Speaking Nearby": A Conversation with Trinh T. Minh-Ha', *Visual Anthropology Review*, 8 (1): 82–91.

Chow, R. (2014), *Not Like a Native Speaker: On Language as a Postcolonial Experience*, New York: Columbia University Press.

Curran Bernard, S. (2015), *Documentary Storytelling: Creative Non-Fiction on Screen*, 4th edn, Waltham, MA: Focal Press.

Elphick, J. (2016), 'Oyster Factory – An Interview with Kazuhiro Soda', *Fourthreelfilm.com*, June 17. Available online: http://fourthreefilm.com/2016/06/oyster-factory-an-interview-with-kazuhiro-soda/ (accessed 28 February 2017).

Fanon, F. (1967), *Black Skins, White Masks*, trans. C. L. Marckmann, New York: Grove.

Fitzsimmons, T. (2009), 'Braided Channels: A Genealogy of the Voice of Documentary', *Studies in Documentary Film*, 3 (2): 131–146.

Honess Roe, A. (2013), *Animated Documentary*, London: Palgrave Macmillan.

Kunreuther, L. (2014), *Voicing Subjects: Public Intimacy and Mediation in Kathmandu*, Los Angeles, CA: University of California Press.

Martineau, B. H. (1984), 'Talking about Our Lives and Experiences: Some Thoughts about Feminism, Documentary, and "Talking Heads"', in T. Waugh (ed.), *'Show Us Life': Toward a History and Aesthetics of the Committed Documentary*, 252–273, Metuchen NJ: Scarecrow Press.

Mayne, J. (2002 [1990]), 'Women in the Avant-Garde: Germaine Dulac, Maya Deren, Agnès Varda, Chantal Ackerman, and Trinh T. Minh-ha', in W. Winston Dixon and G. A. Foster (eds), *Experimental Cinema: The Reader*, 81–112, London: Routledge.

Nichols, B. (1983), 'The Voice of Documentary', *Film Quarterly*, 36 (3): 17–30.

Rangan, P. (2011), 'Immaterial Child Labor: Media Advocacy, Autoethnography, and the Case of *Born into Brothels*', *Camera Obscura*, 25 (3): 142–177.

Rangan, P. (2015), 'In Defense of Voicelessness: The Matter of the Voice and the Films of Leslie Thornton', *Feminist Media Histories*, 1 (3): 95–126.

Renov, M. (2004), *The Subject of Documentary*, Minneapolis, MN: University of Minnesota Press.

Sarkar, B. and J. Walker (2010), 'Introduction: Moving Testimonies', in B. Sarkar and J. Walker (eds), *Documentary Testimonies: Global Archives of Suffering*, 1–34, New York: Routledge.

Soda, K. (2017), Email to the author, 21 February.

Stegewerns, D. (2003), 'The Japanese "Civilization Critics" and the National Identity of their Asian Neighbours, 1918–1932: The Case of Yoshino Sakuzo', in L. Narangoa and R. B. Cribb (eds), *Imperial Japan and National Identities in Asia, 1895–1945*, 107–128, London: Routledge Curzon.

Trinh, T. M.-H. (2016 [1984]), 'Mechanical Eye, Electronic Ear, and the Lure of Authenticity', in J. Kahana (ed.), *The Documentary Film Reader*, 758–762, Oxford: Oxford University Press.

Tsui, C. (2015), '"Oyster Factory" ("Kaki Kouba"): Locarno Review', *Hollywood Reporter*, 13 August. Available online: http://www.hollywoodreporter.com/review/oyster-factory-kaki-kouba-locarno-815089 (accessed 28 February 2017).

Winston, B. (2008), *Claiming the Real II: Documentary: Grierson and Beyond*, London: BFI Publishing.

Young, L. (1999), 'Rethinking Race for Manchukuo: Self and Other in the Colonial Context', in F. Dikötter (ed.), *The Construction of Racial Identities in China and Japan: Historical and Contemporary Perspectives*, 158–176, London: Hurst & Company.

13

Vocal Hierarchy in Documentary

James Nicholson

When we watch a documentary, our engagement with the voices we hear is first a matter of meaning, of understanding what people are saying. Voices carry much of the content of documentaries, whether through narration, interviews or in the social interaction captured by the camera and sound recorders. However, voices are much more than a means to convey information. Voices in documentary are inextricably linked to issues of power. To 'have a voice' in the wider world is in some sense to have power and recognition; similarly, the presence or absence of voices in documentaries grants power to some and denies it to others. The disposition of voices expresses the way in which the filmmaker exercises his or her power over the material of the film. Voices also determine the form of a documentary. The number and identity of speakers, the way they are presented to the audience and the way voices relate to images are important factors in establishing the individual character of every documentary.

Given the centrality of voices to documentary, it seems logical that we should pay close attention to their use and that methods for making a close reading of voices should have been developed. However, so far, this has not been the case. While the voice has not been neglected in academic literature it has been discussed mainly in terms of historical development (Chanan 2007), rhetorical function (Nichols 1991, 2010) and ideological debate (Bruzzi 2000). There has been little systematic analysis of voices in terms of their deployment, the interaction between voices and the interventions that shape and control them in post-production. This chapter establishes some starting points for making a close reading of voices in documentary film. It is a result of several years of work that began with my efforts to clarify Bill Nichols's concept of modes of documentary for students engaged in both documentary studies and production, and led me to Michel Chion's work on sound in fiction films. I found that some of Chion's most important concepts are also applicable to documentary. As a teacher of documentary production, I seek to relate theory to practice, and my research has been informed throughout by my long experience of documentary production, both in the field and in post-production.

Using Werner Herzog's *Grizzly Man* (2005), James Marsh's *Man On Wire* (2008) and Stephen Walker's *Young@Heart* (2007), I will explore the types of voice that are distinctive to documentary, the power relationships between voices and the ways in which the power of a voice is affected by its position in sound space. This analysis shows that the deployment of voices establishes a filmmaker's stance as a narrator who speaks to the audience, an interlocutor who engages in dialogue with his or her subjects or an observer who records the voices of others. I argue that when a filmmaker uses multiple voices the interaction between voices is governed by a hierarchy of power in which narration dominates over interview and overheard voices. This means that the filmmaker's choice to use narration, interview and/or overheard voices is a fundamental factor in the position the filmmaker adopts towards his or her subjects. The filmmaker's control of voices is also a central concern of my analysis. The voice is highly malleable in post-production, and filmmakers can alter the meaning of voices as they position them relative to the images and other voices in the film.

Sound theorist Michel Chion, in *Audio-Vision: Sound on Screen* (1994) and *The Voice in Cinema* (1999), investigates the interaction of sound and image and the perception of sound by the film viewer. He elucidates two important concepts that explain the power of the voice: first, he points out that human hearing is 'vococentric' (Chion 1999: 5). He asserts that for human beings, the voice is the most important sound (Chion 1999: 5) and this attention to the voice is involuntary. This explains why, when watching a film, we attend to the voice before all other sounds. Second, Chion articulates the concept of 'added value' when image and sound work together. Sound does not simply accompany the image. Rather, image and sound are engaged in constant interaction and the soundtrack of a film cannot be considered as an entity in itself:

> Added value works reciprocally. Sound shows us the image differently than what the image shows alone, and the image likewise makes us hear sound differently than if the sound were ringing out in the dark. (Chion 1999: 21)

Chion's writing concentrates on fiction film, but the concepts of added value and vococentric hearing apply equally in other contexts because they are universal human tendencies (Chion 1994: 6; Langjaer 2010). As such, both added value and vococentric hearing are independent of film form or genre, and the concepts can be readily extended to non-fiction film.

Chion's exploration of sound spaces in cinema is also applicable to documentary. A distinctive feature of documentaries is that they make much greater use of non-diegetic voices than fiction films and that voices in documentaries frequently shift between sound spaces. These shifts are significant because they alter the power relationships between voice and image, and between voices.

Bill Nichols establishes the centrality of the voice in his theoretical framework, describing three distinct ways in which voices address the audience:

> Whether it is through what we hear a commentator tell us about the film's subject, what social actors tell us directly via interviews, or what we overhear social actors say among themselves as the camera observes them, documentaries depend heavily on the spoken word. (Nichols 2010: 28)[1]

Here, Nichols refers to three types of voice that are central to documentary: the *direct voice* that addresses the audience – sometimes the voice of an on-screen presenter, but more commonly that of a voice-over narrator; the *indirect voice* which is addressed to the filmmaker, usually in an interview; and the *overheard voice*, where the speech of social actors is directed at neither the audience nor the filmmaker, but at each other, and is recorded seemingly without intervention.

Nichols explicitly links the direct, indirect and overheard voices to three of his documentary modes. Expository documentaries 'take shape around commentary directed toward the viewer' (Nichols 1991: 34). Observational documentaries 'are characterized by speech overheard rather than heard since the social actors engage with one another rather than speak to the camera' (Nichols 1991: 39). In the participatory mode, which is characterized by the use of interviews, the filmmaker 'recruited' social actors' 'comments and responses provide a central part of the film's argument' (Nichols 1991: 44). Nichols thus recognizes that the direct, indirect and overheard voices address the audience in different ways that mark the distinction between the expository, participatory and observational modes. In doing so, he recognizes that the disposition of voices establishes the relationships between filmmaker, subject and audience.[2] Some documentaries use only one type of voice: *March of the Penguins* (Luc Jacquet 2005), for example, uses only the direct voice of the narrator, while *Don't Look Back* (D. A. Pennebacker 1965) uses only overheard voices. However, many documentaries, including *Grizzly Man* and *Young@Heart*, use combinations of direct, indirect and overheard voices and contain elements of more than one of Nichols's modes.

A useful starting point for analysis of a particular documentary is to identify the combination of direct, indirect and overheard voices it uses. The filmmaker's choices establish a distinct stance. By using the direct voice to address the audience, the filmmaker occupies the role of narrator. By interacting with social actors and eliciting their speech – the indirect voice – the filmmaker assumes the role of a conduit between the subjects and the audience. When the filmmaker forgoes both narration and interaction and uses only the overheard voice, he or she takes up the stance of an observer who represents events as he or she has witnessed them. This stance is a major element of what Nichols terms the 'voice of the documentary' (Nichols 1983), a phrase he uses to describe the authorial perspective and tone of the film. Nichols is at pains to establish that the 'voice of the documentary' encompasses more than the spoken voice, as it is 'formed by the unique interaction of all of a film's codes' (Nichols 1983: 260–261), including cinematography and editing. In spite of this, his choice of the word 'voice' to describe the informing presence of the documentary author is significant. In fact the way voices are used is a significant factor in establishing the documentary maker's perspective and position in the text.

The hierarchy of voices

Of course, many documentaries contain more than one type of voice. In these cases, the different voices interact, establishing a hierarchy of authority.

At the top is the direct voice, especially in the form of voice-over. The authority of the voice-over has been discussed by various theorists and Stella Bruzzi (2000: 40–41) summarizes several when she states that the use of voice-over has been condemned as 'didactic and undemocratic'. Bruzzi sees this as a reaction to the 'voice of God' style of narration, which combines an anonymous narrator, a deep masculine voice (and, increasingly, a celebrity voice, as Annabelle Honess Roe argues in this volume) and an attitude of unquestioned authority. Bruzzi discusses the use of female voices and less formal delivery styles as alternatives to the 'voice of God' approach. However, the direct voice remains at the top of the hierarchy even when it does not conform to the 'voice of God' model, because it enjoys certain advantages over other voices. Voice-over commentary is usually written during post-production, so that the narrator enjoys the position of having a perspective on the completed film; it is recorded at high quality and often performed by a professionally trained speaker; and as such it represents the voice of the filmmaker, whether this is literally true or not.

Even more importantly, voice-over takes advantage of the phenomena described by Chion: the vococentric hearing of humans and the added-value relationship between image and sound. As Chion points out, when a voice is present in the soundtrack, our attention to it is involuntary. Not only do we listen to and try to understand the voice, but we instantly relate it to the images on screen. This means that the direct voice modifies our perception of images by supplying information and expressing emotional attitudes towards them. It can impose cohesion on disparate images and link scenes together. When other voices are present, the direct voice can identify, validate or question these voices, influencing the audience's reception of them. Mary Anne Doane (1985) places voice-over between the audience and the image, saying that it establishes a 'complicity between itself and the spectator – together they understand and thus place the image' (1985: 168). Doane's point applies equally well to voices: the voice-over has the ability to 'place' other voices, as it seems to position itself between the viewer and all other voices in the film.

If the direct voice is entirely absent, the indirect voice can become the dominant voice of the film. In films that use indirect and overheard voices, but have no voice-over, the indirect voice has greater authority than the overheard voice. An interview is usually a purposeful dialogue, researched and steered by the interviewer, so it can articulate an argument more clearly than the overheard voice. The interview voice can also be used to comment on observational scenes, 'placing' the overheard voice. The overheard voice, on the other hand, is rarely used to comment on or contextualize other voices. The overheard voice is not written by the filmmaker, as the direct voice is, nor steered by the interviewer, like the indirect voice. In addition, the overheard voice is often recorded under challenging conditions. The result is that scenes containing only overheard voices tend to yield a partial and inconclusive representation of events, compared with interviews and commentaries, which can take advantage of hindsight, discussion and preparation.

The hierarchy that governs the direct, indirect and overheard voices is evident in documentaries that use multiple voices, such as the award-winning British production, *Young@Heart*. The film follows a group of elderly American singers, famous for their performances of modern rock songs, as they prepare for a major concert. The director, Stephen Walker, performs the voice-over himself, personalizing it by speaking in the first person and by a brief appearance early in the film. Although he avoids the anonymous 'voice of God' style, Walker's is still the voice that controls the film, introducing interviews with background information and providing links between scenes. Observational scenes are often introduced with voice-over and followed by interviews that comment on what we have just seen. Thus, the overheard voices of the observational scenes are subject to comment by both direct and indirect voices.

In contrast, James Marsh's 2008 documentary *Man on Wire* has no direct voice. It tells its story through interviews. The film documents Philippe Petit's ultimately successful attempt to walk a tightrope between the twin towers of the New York World Trade Center in 1974. The narrative is carried by interviews with Petit and his team of collaborators. The interviewees are seen in formal interview set-ups, and their voices are also paired with both archival footage and dramatized re-enactments. There is limited use of overheard voices in the archival footage, and the hierarchy of voices is evident when the interview voices comment on both archival and dramatized scenes, placing the overheard voices in context. Marsh does not give himself any role in the film – all interview questions have been edited out of the final cut. Nichols asserts that in films dominated by interviews, 'textual authority shifts towards the social actors recruited' (1991: 44). This is undoubtedly true in the case of *Man on Wire*, in which Philippe Petit and his companions occupy the centre stage.

Werner Herzog takes the opposite approach in *Grizzly Man*, in which he performs the voice-over himself. The film tells the story of Timothy Treadwell, a self-styled environmentalist who spent thirteen summers camping among the grizzly bears in the Katmai National Park in Alaska. It makes extensive use of video shot by Treadwell and focuses on the death of Treadwell and his girlfriend Amie Huguenard, who were both killed by a bear in 2003. *Grizzly Man* features two direct voices: the voice-over performed by Herzog himself, and Timothy Treadwell's direct address to the camera in numerous clips from his video footage. There is a continual interaction between these two direct voices. Herzog at times interrupts Treadwell to comment on or disagree with him, emphasizing the dominant position of Herzog's voice-over. For example, when Treadwell finds a fox cub that has been killed by wolves, he mourns over the corpse, saying: 'I love you, and I don't understand. It's a painful world.' Herzog breaks in: 'Here, I differ with Treadwell. He seemed to ignore the fact that in nature, there are predators. I believe the common denominator of the universe is not harmony, but chaos, hostility and murder.' These critical comments are balanced by Herzog's praise for Treadwell, especially as a filmmaker: 'He captured such glorious improvised moments, the likes of which the studio directors, with their union crews, can never dream of.' In addition to making these direct statements, Herzog exposes Treadwell's production methods by including parts of Treadwell's prepared speeches where he breaks character between takes or tries different versions of a piece to camera. When Treadwell made these recordings, he seemed to be building a documentary in which his would be the central voice. Both Herzog's commentary on Treadwell's speeches and the revelation of their construction serve to 'place' Treadwell's direct voice as part of the subject matter, rather than the dominant voice of the film. Herzog's assertions of dominance reflect the shift in Treadwell's position from filmmaker to subject.

Both *Young@Heart* and *Grizzly Man* take advantage of the direct voice's power to introduce a degree of reflexivity into the documentary by referring to the process of production. In *Young@Heart*, this is fairly straightforward, with Walker beginning a sequence by telling the audience, in the present tense, where he is going next and whom he will be talking to. These references to the process of making the film do not reveal much of Walker's life nor tell us much about the film's production. Instead, they place Walker in the present tense of the story to encourage audience engagement, even though the voice-over was actually written and performed well after the shoot.

In *Grizzly Man*, on the other hand, Herzog makes more personal references to his own career as a director and also makes it clear when his interjections are a matter of personal opinion. Herzog also reveals that he did not gain access to Treadwell's last tape until late in post-production and includes a scene in which Jewel Palovak allows him to listen to the audio recording of the attack in which Treadwell and Huguenard died. This scene stands out because it is the only scene in which Herzog appears on screen in the film, and it appears to be a piece of unplanned observational shooting. The camera is in position for an interview shot of Palovak, but the operator has zoomed out and reframed to show Herzog from a back-three-quarters angle as he listens to the tape through headphones. Herzog's voice-over introduces the scene as he listens to the recording. There follows a conversation in which Herzog tells Palovak to destroy the recording without listening to it. The dialogue is recorded as overheard speech.

This scene takes the audience behind the scenes of the production, as Herzog temporarily gives up the role of the narrator and instead appears as the director grappling with the story he is trying to tell. Rather than asking Palovak questions from off-screen, he speaks to her as a friend and holds her hand. Johnson

(2008) identifies this as a pivotal scene. He asserts that its impact derives from 'our perception that what Herzog is listening to … is *real*' (Johnson 2008: 70). The observational style of the scene reinforces this with its realism. The scene emphasizes the existence of the recording of the attack, while also showing that Herzog will deny the audience the chance to hear it. Such reflexive moments frame the audience's view of the production process as well as the subject matter and establish the narrator as an individual personality with a distinct viewpoint.

Looking at the balance between direct, indirect and overheard voices, and the hierarchy established by their interaction, is a useful way to start analysing a documentary's use of voices. It brings into focus the stance the filmmaker has chosen – narrator, interlocutor, observer or some combination of these. It also reveals the power structure of the film. *Grizzly Man* shows how the introduction of a voice-over can diminish the authority of voices contained in archival footage. The imposition of Herzog's narration deposes Treadwell's voice from its intended place as the dominant voice of the film. In *Man on Wire*, the absence of voice-over has a marked effect: the voices of social actors come to the fore, and the filmmaker's engagement with them becomes less obvious because the interviewer's voice is also absent. In *Young@Heart*, multiple voices are used, and the filmmaker's stance may seem to change from moment to moment. But the voice-over positions Walker as a narrator and influences the audience's reception of all other voices in the film. The numerous interviews are characteristic of the participatory mode, and for some viewers, the sense of encounter with the chorus members will be the foremost impression they take from the film. The observational scenes are also very important, particularly those that depict live performances, which are strongly cathartic and provide the climax of the film. But although *Young@Heart* makes use of the strategies of interaction and observation, Walker's voice is a constant presence. He introduces and comments on observational scenes and he identifies and describes all the interviewees, thereby 'placing' all other voices in the film and asserting his power as the dominant voice.

Sound spaces

Another way of approaching voices in documentary is to examine their placement in sound spaces. Both Michel Chion (1994: 73) and Michael Chanan (2007: 115) identify three cinematic sound spaces: the on-screen, the off-screen and the non-diegetic. When the source of a sound is visible on screen, it is in the on-screen space. When the source of the sound is not visible, but is nevertheless a part of the scene, it is in the off-screen space. Sounds in the on-screen and off-screen spaces are termed diegetic because their source is located in the scene, even if the source is not visible. The non-diegetic sound space is the domain of sound whose source 'is not only absent from the screen, but is external to the story world' (Chion 1994: 73). The source of a non-diegetic sound cannot be seen within the frame, nor inferred as an off-screen part of the action.

In documentary, the direct voice may be used in the on-screen space, as it is by Timothy Treadwell in the clips used by Herzog, but more commonly the direct voice is placed in the non-diegetic sound space in the form of voice-over. The indirect voice is often an on-screen sound, as it is in *Grizzly Man*, where interview voices are always accompanied by synchronized images. But the indirect voice is also frequently used in the non-diegetic sound space, as it is in *Man on Wire*, where interview voices speak over archive footage and re-enactments. The overheard voice, by contrast, is essentially a diegetic sound, and is nearly always used in the on-screen and off-screen sound spaces. Observational documentaries, such as *Don't Look Back* (D. A. Pennebaker 1965), *High School* (Frederick Wiseman 1968) and *To Be and to Have* (Nicolas Philibert 2002), avoid the use of non-diegetic voices altogether. In documentary, the placement of voices in sound space is important because when voices move from the on-screen space to off-screen or non-diegetic space, they often gain power. This applies not only to the direct voice, but also to interview and overheard voices.

Young@Heart contains a scene that places particular emphasis on the overheard voice by placing it in the off-screen sound space. This increases the audience's engagement with the speakers, in spite of their invisibility. The singers have just boarded a bus, which is to take them to stage a performance at a local prison. The voice-over explains that Diane, the chorus director's assistant, has asked the film crew to stay off the bus while she makes an announcement to the singers about the death of one of their members, Bob Salvini. However, it appears someone on the bus is wearing a radio microphone, as the announcement is recorded with perfect clarity. The sound of the announcement and the chorus members' reactions is laid under a long shot of the exterior of the bus. The combination of limited visual content and close-microphone sound throws more emphasis on the overheard voices. In spite of the extreme contrast between image and sound perspective, this construction enhances the realism of the scene, because it promotes the impression that an unguarded, spontaneous moment of conversation has been captured. The moment emphasizes the sense of being in the present, as the audience hears the news at the same time as Salvini's fellow singers. The scene is a powerful demonstration of the emphasis that can be placed on the voice by using it in the off-screen sound space.

Non-diegetic sound space

In documentary, the non-diegetic sound space is the most powerful position for the voice. Voice-over has power over the image because of the added-value effect, and as Doane points out, it establishes a position of complicity with the viewer (Doane 1985: 168). *Grizzly Man* shows how Herzog's non-diegetic voice-over dominates Treadwell's on-screen voice, even though both voices address the audience directly. Interview voices can also move from the on-screen sound space to the non-diegetic. The meaning of both image and sound is altered when this happens, because the interview voice gains power from the shift.

In *Man on Wire,* Marsh uses formal interview set-ups, but also pairs interview voices with archival footage and re-enactments, placing the voice in the non-diegetic sound space. His interview voices sometimes migrate from on-screen to non-diegetic space in mid-sentence as he cuts from an interview shot to archive footage or re-enactment. This is a familiar technique to viewers, so familiar that we may not stop to reflect that whenever an interview voice moves into the non-diegetic sound space, its meaning is altered as its relationship to the pictures changes, because of the reciprocal relationship between image and sound that Chion terms 'added value' (Chion 1999: 21). For example, when Petit's girlfriend, Annie Allix, describes their relationship, archival footage shows her watching Petit as he practises on a wire set up in his garden. The original sound is kept low in the sound mix, so that the interview voice holds centre stage. Allix, who has already been seen in an interview early in the film, is now seen as a young woman. She describes herself as 'painfully shy' and 'overwhelmed' by Petit's pursuit of her. The archival shots that accompany this show Petit active, walking and running on the wire set up in his garden, while Allix stands below, watching him. After a pause in the voice track, the on-screen action changes. Allix is now with Petit on the wire, walking behind him, her hands on his shoulders. Her interview voice track resumes:

> We became inseparable. In fact, my life was completely consumed by his, and he never thought to ask me whether I had my own destiny to follow. It was quite clear I had to follow his.

Because of the added-value effect, Allix's voice seems to tell the audience how to see the images. The first part of the sequence appears to illustrate Allix's account of their relationship: Petit is active, Allix is the passive observer; Petit is seen in low-angle shots, Allix in a high angle from the wire or from a neutral angle. The second section, showing Allix holding on to Petit as she crosses the wire behind him, also presents a striking image when accompanied by her words: they are 'inseparable' and the line of the wire stands in for Petit's destiny, which they must both follow.

This is an example of the power of the added-value effect, as the combination of footage and voice in this scene has altered the meaning of both elements – the voice seems to be commenting on the footage, while the imagery seems to be offering proof of Allix's statements. The words and images seem to fit so well that it is easy to forget that this construct is the editor's creation. The scene was shot about thirty-five years before the interview was recorded, and the interpretation that Marsh has placed on the scene so many years later could not have been foreseen at the time of its filming. Furthermore, when Allix was interviewed, she was not necessarily thinking of this footage. In fact, she may never have seen it. Yet laying her voice track under these images suggests that this is what she is thinking of as she speaks, influencing our interpretation of her words.

Placing the interview voice in the non-diegetic sound space in a sustained sequence gives the voice some of the power of a voice-over commentary, but the effect is not exactly the same. In the garden scene, Allix is visible as her younger self, and this reminds the viewer that her voice is not that of a detached narrator, but that of a social actor. This use of the indirect voice is often referred to by directors as 'thought track', a term that expresses the idea of an interior voice, rather than public address. The sense of the interior and intimate is strongly present when Marsh pairs Allix's voice with the archival footage of the garden scene, distinguishing it from the more detached voice of a narrator.

Marsh's placement of interview voices in non-diegetic sound space contrasts with the tactics used by Herzog and Walker. In *Young@Heart* and *Grizzly Man*, interviews always show the interviewee in shot, keeping the voice in the on-screen sound space. Both Herzog and Walker can be heard asking questions from off-screen. Although they are not seen in the majority of interviews, their voices in the off-screen sound space establish their presence in the scene. In

Man on Wire, by contrast, the interviewer has been effaced, and is neither heard nor seen.

The use of sound spaces reflects the power structure in each film. In *Young@Heart* and *Grizzly Man*, the most powerful sound space, the non-diegetic, is reserved for the director's voice-over. In *Man on Wire*, Marsh cedes this space to the interview voices, increasing their authority further by eliminating the interviewer's voice and forgoing the use of voice-over. Herzog and Walker use sound space to maintain their authorial presence in the film, not only through the non-diegetic voice-over, but also by using the off-screen sound space during interviews. Marsh pursues the opposite course, and by his absence from the soundtrack, cedes textual authority to the interviewees.

Editing

It is necessary to focus on editing when considering the voice in documentary, because of the control filmmakers have over voices in the editing process. This applies particularly to a film such as *Man on Wire*, because of its foregrounding of the interview voice. While the story is told in the voices of social actors, their apparent authority cannot be taken at face value, because of the filmmaker's hidden intervention. The creation of a coherent voice track is often the first concern of the documentary editor, because many interviews and overheard voice recordings are not coherent and concise enough for the filmmaker's purpose. Editing voice tracks makes speakers more articulate and entertaining, but it inevitably alters the meaning of the original speech.

When the interview voice is in the non-diegetic sound space, as in the garden scene in *Man on Wire*, the opportunities for editing the voice are increased, because the editor does not have to make picture cuts to conceal edits in the voice track. This gives the editor the opportunity to engage with the voice on a much more detailed level. In recent years, digital audio editing tools have increased the filmmaker's power in post-production, while also making it easier to conceal the exercise of this power over the voice. Editors can construct new sentences and phrases, sometimes by making edits in the middle of a word.

Sound editor John Purcell (2013) explains that it is common practice for documentary directors to transcribe an interview, then construct an edit by cutting and pasting the transcript in a word processor (Purcell 2013: 338–339). This text-based edit, made up from disparate pieces of the interview, does not take account of what the result will actually sound like. The dialogue editor encounters two main challenges: 'cadence problems caused by assembling sentences from different parts of an interview' and 'incorrect terminations caused by a sentence chopped off before the phrase actually ends, resulting in an unconvincing ending inflection' (Purcell 2013: 338). Sound editors have techniques for dealing with these unnatural-sounding cadences and inflections, such as substituting a word for one with a different inflection; combining the beginning of one word with the end of another; and changing speed and/or pitch to alter the inflection of a word or the cadence of a phrase or sentence (Purcell 2013: 339–341).

This means that editors can create grammatical constructions that were not in the original recording, fundamentally altering the meaning of the interviewee's words. Therefore, although interviews appear to allow the interviewees to tell their own story in their own words, in many cases the meaning of the interview voice has been modified by extensive editing. We cannot know from the finished film the extent of this intervention. This suggests that what Nichols terms the 'textual authority' of social actors in films like *Man on Wire* cannot be taken at face value, because the film's construction disguises the extent of the filmmaker's control. The interview does not arrive in the finished film in its original form, but as an interpreted representation of the original.

However, the filmmaker does not enjoy absolute control when editing interview voices. Even with modern technology, the editor is limited by the material. Ethical and legal considerations deter filmmakers from distorting an interview's meaning. The time and money available to the filmmaker can be limiting factors, too. The production budget may limit or preclude the employment of highly skilled dialogue editors. Nonetheless, when analysing films like *Man on Wire* that use non-diegetic interview voices, we are left with a large area of uncertainty about the extent to which the filmmaker has reinterpreted the interview material.

Conclusion

The voice is a key element through which we can analyse the power relations between filmmaker, subjects and audience in documentaries. The hierarchy of power that operates between direct, indirect and overheard voices means that the choice to use or exclude any of these voice types has a crucial impact on the film. By means of these choices, directors position themselves in the film, choosing whether to address the audience directly, as Herzog and Walker do, or to remain unseen and unheard, as Marsh does. The use of sound spaces and the editing of voices also involve stylistic choices that reflect the exercise of power, whether hidden or overt.

The approach I have suggested here can be applied to any documentary, and may be useful whether the voice is the central concern of the scholar or not. The interactive relationship between sound and image means that a focus on the spoken word can contribute to the reading of images, text and editing technique. A reading of the voice is a most effective way to discern the authorial stamp that Nichols (1983) terms the 'voice of the documentary'.

Notes

1 Nichols here and elsewhere uses the term 'social actors' to mean people who appear as themselves in interviews and actuality scenes in documentaries.

2 The other modes Nichols defines, the poetic, the reflexive and the performative, are not defined by the use of a particular voice type in the way that the expository, observational and participatory modes are. However, the way voices in these modes are positioned also has a fundamental influence on the style of the film.

References

Bruzzi, S. (2000), *New Documentary: A Critical Introduction*, London: Routledge.

Chanan, M. (2007), *The Politics of Documentary*, London: British Film Institute.

Chion, M. (1994), *Audio-Vision: Sound on Screen*, New York: Columbia University Press.

Chion, M. (1999), *The Voice in Cinema*, New York: Columbia University Press.

Doane, M. (1985), 'The Voice in Cinema: The Articulation of Body and Space', in E. Weis and J. Belton (eds), *Film Sound: Theory and Practice*, 162–176, New York: Columbia University Press.

Don't Look Back (1965), [Film] Dir. D. A. Pennebaker, USA: Leacock-Pennebaker.

Grizzly Man (2005), [Film] Dir. Werner Herzog, USA: Lionsgate Films, Discovery Docs.

High School (1968), [Film] Dir. Frederick Wiseman, USA: Osti Productions.

Johnson, D. (2008), 'You Must Never Listen to This': Lessons on Sound, Cinema, and Mortality from Herzog's 'Grizzly Man', *Film Criticism*, 32 (3): 68–82.

Lankjaer, B. (2010), Making Fictions Sound Real – on Film Sound, Perceptual Realism and Genre, *Mediekultur*, 48: 5–17.

Man on Wire (2008), [Film] Dir. James Marsh, UK: Icon Productions, Magnolia Pictures.

Nichols, B. (1983), The Voice of Documentary, *Film Quarterly*, 36 (3): 17–30.

Nichols, B. (1991), *Representing Reality: Issues and Concepts in Documentary*, Bloomington, IN: Indiana University Press.

Nichols, B. (2010), *Introduction to Documentary*, Bloomington and Indianapolis, IN: Indiana University Press.

Purcell, J. (2013), *Dialogue Editing for Motion Pictures: A Guide to the Invisible Art*, New York and London: Focal Press.

Young@Heart (2008), [Film] Dir. Stephen Walker, UK: Fox Searchlight Pictures, George Walker Films, Channel 4.

Section Five

Time, Testimony and History

14

Crime Documentary's Confessing Voice

Kristen Fuhs

> *People who are innocent don't confess. The defendant confessed because he was guilty, because he did it.*

These words, spoken during prosecuting attorney Tom Fallon's closing arguments, emerge in the ninth episode of the popular Netflix documentary series, *Making a Murderer* (Moira Demos and Lucia Ricciardi 2015). The documentary series focuses on the prosecution of recent exoneree Steven Avery, and his conviction for a second crime he may not have committed; however, it is his nephew, Brendan Dassey, who emerges as the series' most tragic figure. Sixteen years old at the time of his arrest, with an IQ that is significantly below average, Dassey becomes a pawn in the prosecution's quest to implicate Avery for the murder and dismemberment of Teresa Halbach. Questioned over the course of many hours, on four separate occasions, without parental or legal supervision, Dassey is coaxed into confessing his role as an accomplice to the grisly crime. Archival footage from his first police interrogation reveals Dassey to be quietly withdrawn, confused and highly suggestible; detectives prompt him to recall events, but often resort to feeding him details when his words are inaccurate or slow to come, even as they prod him to 'be honest' and 'tell the truth' about his involvement. Throughout the interrogation, the tenor of his voice – reticent, hesitant, uncertain – contradicts at every point the narrative produced by his words: but it is the words, not the voice, that will ultimately become evidence used to bolster the case against him.

This is not the only fraught confession we see coaxed from Brendan in *Making a Murderer*. His court-appointed attorney, Len Kachinsky, hopes to secure him a plea bargain and enlists his investigator, Michael O'Kelly, to get a statement from Brendan that will aid in this endeavour. O'Kelly gives Brendan a form to fill out – not an official legal document, but rather something Kachinsky drew up for the purposes of this interview – that asks Brendan not whether he's guilty, but if he's sorry, and then to outline the reasons why. Brendan has trouble with this form – after all, he can't apologize for something he didn't do – so he writes down his own memories of what happened on the day in question, which have nothing to do with the murder of Teresa Halbach. O'Kelly rejects this statement as false, however, and has Brendan start over on a new, more 'truthful' statement. Whenever Brendan pauses in his writing, O'Kelly guides him: 'Why don't you draw another picture over here, of him stabbing her … Why don't you draw another picture right here, of you having sex with her … Why don't you draw a picture of the bed and how she was tied down, big-sized so we can see it.'

Brendan is almost completely voiceless in this scene, and when he does speak, it is through a series of mumbled, monosyllabic words. In fact, he is so quiet, his words so indistinct, that the filmmakers provide subtitles to clarify what he is saying. His body language further signifies his lack of agency – slouched in his chair, eyes cast down, he seems to draw more into himself as the story veers further and further away from his own recollections. Michel Foucault, writing about the relationship between justice and the act of confession, has suggested that confession 'is a verbal act through which the subject affirms who he is [and] binds himself to this truth', but the final statement O'Kelly elicits from Brendan has neither this spoken clarity of purpose nor this attachment to authenticity (Foucault 2014: 17). Why couldn't – or wouldn't – O'Kelly hear in Brendan's voice, or recognize in his voicelessness, the truth that seems so clear to me as an observer of this interaction?

This particular confession is never used as direct evidence against Brendan in court.[1] I highlight it here, however, because this is the film's clearest illustration of the way in which authority figures who surround Brendan – the police, his attorneys, this investigator, even his mother and uncle – manipulate his statements by shaping which of his narratives are deemed valid as 'true' accounts. This conversation between Brendan and O'Kelly – along with the three other recorded admissions of guilt he makes in more official settings – highlights more than anything else the extent to which the act of confession is a ritual grounded in relations of power and knowledge. It is because of this exercise of power relations, Foucault reminds us, that all confessions are 'costly' to the individuals caught up in the exchange (Foucault 2014: 17). So, how could the courts – and how can we as viewers of this documentary – possibly judge the veracity of Brendan Dassey's confession when he speaks through such a disempowered voice?

This chapter will examine the rhetorical power of confession in the contemporary US crime documentary by focusing on the extent to which the subject's voice acts as a marker of authenticity and veracity. As many scholars have argued, confession has become a dominant form of self-examination and self-expression in contemporary society (see Brooks 2000, 2005; Diamond 2010; Payne 2008; Tell 2012). Despite its non-secular roots as a private act of contrition, we now live in an era of conspicuous public confession: from talk shows, to reality television, to the personal scandals of public leaders, 'confession has, arguably, never played such a large role in political and social life' (Payne 2008: 24).[2] And yet despite, or even perhaps because of, the ubiquity of confession in this contemporary cultural moment, confession's status as a signifier of truth has been increasingly called into question. In fact, asks the legal scholar Peter Brooks, if confession comes so easily, shouldn't we be sceptical about whether there's any truth value at all in these utterances? (Brooks 2005: 74) Of course, there are many reasons why people confess: guilt, vengeance, coercion, justification, mental impairment, absolution, self-abasement. Confession can serve any number of motives, and telling the truth isn't necessarily one of them.

Making a Murderer and *The Jinx* (Andrew Jarecki 2015) are two recent multi-episode US documentaries that throw this confusion about the truth value of confessional speech into stark relief. *The Jinx* investigates the life and crimes of the accused serial murderer Robert Durst, whose seemingly accidental admission of guilt in the closing minutes of the documentary's final episode made *Jinx* the stuff of pop cultural catnip in the weeks following its initial broadcast. Unsettled and breathless, Durst belches his way through a speech that seems to avow his criminality, ending with a quizzical: 'What the hell did I do? Killed them all of course.' Recorded off-camera, over a still-hot mic, this disembodied utterance begs the question: Where do we locate the evidential properties of the human voice as we try to discern how and when speech rises to the level of proof? Although the confessional quandary generated by *Making a Murderer* may not be as philosophically uncertain as that in *The Jinx*, it is no less troubling for its elicitation. In both series, the particularities of the confessing voice call the evidentiary status of documentary speech into question by highlighting the importance of the voice as a medium that carries with it its own evidentiary properties. In other words, when it comes to interpreting confessional speech, analysing the specific aural qualities of the speaking subject – how these confessions are expressed socially and sonically – is just as significant as listening to the words that have been spoken. As such, these series highlight just how crucial the context and conditions of confession are in this particular moment, both in documentary and in culture at large.

Confession: Queen of all proofs?

In US courts, confession is commonly treated as the 'queen of all proofs' (Brooks 2000: 4). Classic texts, such as Charles McCormick's *Handbook of the Law of Evidence*, even suggest that 'the introduction of a confession makes the other aspects of a trial in court superfluous' (1972: 316). However, as the psychology professor and legal expert Saul Kassin has persuasively argued, 'confessions are fallible' (2008: 1309). Statistics belie the belief that 'people who are innocent don't confess,' despite Tom Fallon's assertion in *Making a Murderer.* In fact, 'compelled false confessions are not only possible, but happen with unsettling regularity' (Lassiter et al. 2011: 111). The Innocence Project indicates that of the more than 350 people who have been exonerated through post-conviction DNA testing in the United States, more than a quarter of them had given false confessions (2017). Research points to a number of causes for false confessions: aggressive or deceptive interrogation techniques, a suspect's mental or physical impairment, fear of violence or the actual infliction of harm and failures of due process. Juveniles and the mentally ill are especially vulnerable, because they are particularly susceptible to suggestion and are often eager to please authority figures (Shipler 2012: 53).

Over the past decade or so, a number of documentary films have troubled the associative link between confession and truth telling, highlighting the act's inherent complexities while casting doubt on its assumed evidentiary properties. Like *Making a Murderer*, many of these films are encounters with false or coerced confessions. *The Central Park Five* (Ken Burns 2012), for example, details the process by which five teenagers were made to confess to the rape and attempted murder of a jogger in New York's Central Park in 1989. Also released in 2012, *Scenes of a Crime* (Grover Babcock and Blue Hadaegh) takes prevailing interrogation techniques to task by juxtaposing a police training video with a tape of these techniques in action, where the end result is a man confessing to the abuse and murder of his infant son, even as doctors later rule the child's death was caused by a viral infection. *The Confessions* (Ofra Bikel 2010) follows the story of the Norfolk Four, a group of young Navy veterans who were convicted for rape and murder even though all other evidence in the case contradicted their confessions. *A Murder in the Park* (Christopher Rech and Brandon Kimber 2015) turns the spotlight back on innocence investigations by profiling a case in which a crusading journalism professor and his students seem to have coerced a false confession from their alternative suspect, thereby sending an innocent man to jail as part of their campaign to overturn a questionable conviction.

Documentaries like the ones mentioned above showcase the ease by which confession can lead to miscarried justice, thereby troubling the confessional speech's value – and its legitimacy – as a form of proof. Delivering the majority opinion in *Oregon v. Elstad*, a case that considered the boundaries of 'voluntary' confession, US Supreme Court justice Sandra Day O'Connor argued that because confessions may stem from any number of factors, 'It is difficult to tell with certainty what motivates a suspect to speak' (1985). And yet, as viewers of these documentary films, we are put in a position to analyse just that. Consider, for example, the final minutes of Errol Morris's persuasive account of justice gone awry, *The Thin Blue Line* (1988). Billed as 'the first movie mystery to actually solve a murder', *The Thin Blue Line* used re-enactments to reveal discrepancies in the witness statements that had led to the wrongful conviction of Randal Dale Adams. Over the course of the film one witness – a former teenage runaway named David Harris – emerges as the crime's most likely culprit. Morris's final interview with Harris ends the film, and this encounter seems to affirm Harris's culpability when he admits that he's 'the one who knows' that Randall Adams didn't kill Officer Robert Wood.

How are we to make sense of this exchange between subject and filmmaker? Is Harris actually confessing to the crime here, or is this merely an admission to the role he played in framing Adams? Focusing solely on the words uttered during an on-screen confession can only get us so far in our interpretation of a confession's linguistic meaning. As Lawrence Kramer argues:

> The voice of the subject, the subject as voiced, comes to know itself (insofar as it can ever know itself in words), in the language it speaks, the vocabularies it calls on, the shapes it imparts to its utterances, the habits of its speech acts. At the same time – literally, always: at the same time – voice presents itself to the other as timbre, as intonation, as rhythm. (Kramer 2014: vii)

In addition to the spoken word, a confessor's tone, intonation, pitch, volume, pacing, even his moments of silence are all signifiers of meaning. A mumbled confession conveys something distinct from one that is shouted; a clear tone can suggest something different from one that is full of vocal modulations.

Consequently, the tonal qualities of the voice can act as their own evidence, related to and yet distinct from the intended meaning of any spoken words.

This final interview between Morris and Harris in *The Thin Blue Line* was captured via audio recording – there is no video of this exchange, so we cannot see Harris's body language or his facial expressions when he explains his role in the affair. Instead, we are left searching for vocal clues to contextualize the meaning of his words. Harris has a slow, southern drawl, but the pace of his delivery is deliberate, not evasive. There is a wryness to his voice, which suggests a coy approach to the interview – he's not dissembling, but he's not being entirely forthcoming either. His modulated tone makes him sound forthright, but his periodic chuckles hint at a playfulness, as if he's enjoying his control over the known facts of this narrative. This point is made clear when Morris asks Harris if he was surprised that the police blamed Adams for Officer Wood's death, and he quickly responds, 'They didn't blame him. I did'. The tone, pacing and timbre of Harris's voice throughout the interview lend credence to the substance of his words, allowing for an easy interpretation of this speech as an act of voluntary confession.[3]

In contrast, Brendan Dassey's vocal expression in the first of his official recorded police interviews in *Making a Murderer* seems to contradict the connotation of his words. Two hours into the interrogation, detectives have finished laying out their version of Teresa Halbach's murder. Throughout the interview, Brendan has either agreed with them out of rote reflex ('uh huh') or speculated wildly about the crime's details – Brendan's answers to their questions almost always end with a question mark, his rising intonation a clear signal that he lacks conviction in what he is saying. He mumbles indistinct answers to questions, and there are long pauses between his responses, as if he's just searching for the one thing he can say that will bring this whole encounter to a close. Brendan's body language in this scene further reinforces the hesitance in his speech. Throughout the interrogation, he slouches in his seat, his body just slightly too large for the room's small sofa (see Figure 14.1). His hands are stuffed into the pockets of his baggy stonewashed cargo jeans or folded in his lap, his eyes downcast as he responds to the detectives' questions. Brendan's voice and the visuals align to produce a persuasive image of disempowerment; Brendan is not in a position to contradict the narrative of guilt that police detectives brought with them into the interrogation.

Distressingly, Brendan does not seem to understand the precariousness of his position, and this becomes clear at the end of the interview when he asks detectives

Figure 14.1 Brendan Dassey interrogation tape, *Making a Murderer* (Moira Demos and Lucia Ricciardi 2015).

if they will be finished in time for him to return to school so he can hand in a class project. In stark contrast to the statements delivered during his interrogation, Brendan's voice is animated and hopeful when he poses this question. He sits up, he makes eye contact with the detectives and there is a guilelessness to his tone that makes clear he doesn't understand the associated consequences of the confession that has just been elicited: he has done what's been asked of him, and now he'd like to return to his regularly scheduled programme. This is Brendan's first assertive moment in the entire interrogation, and it has nothing to do with Teresa Halbach's murder. His voice and demeanour here contrast with those of his previous utterances and serve to underscore just how uncertain and hollow the words that would amount to his official confession sounded.

What becomes clear from this example is that Brendan's voice has a materiality all its own, distinct from the words he is speaking, and this voice carries with it its own evidential properties that signify meaning. In discussing the voice in cinema, Michel Chion states, 'From the speech act, we usually retain only the signification it bears, forgetting the medium of the voice itself' (1999: 1). Likewise, Mladen Dolar suggests that 'what defines the voice as special among the infinite array of acoustic phenomena, is its inner relationship with meaning'. This 'intimate connection with meaning', he argues, means the voice is endowed 'with an inner intentionality' and a 'subjectivity which "expresses itself" and itself inhabits the means of expression' (2006: 14–15). In other words, the voice is not merely a vehicle for the production of speech. Rather, the voice is a manifestation of the subjective expression of the self. It makes possible the delivery of the spoken word, but it also conveys its own, distinct meaning.

Focusing on the voice as medium offers a through line that engenders a direct connection between *The Jinx* and *The Thin Blue Line*. Both films make use of audio-taped confessions, and there is a clear accord between their final scenes, which pair the respective recorded confessions with a minimalist mise en scène. David Harris's boastings play over a series of static close-ups of a mini cassette recorder; Robert Durst's play over the recently vacated hotel conference room that had been the site of their final interview. There is little of visual interest in either scene – and this encourages viewers to focus their attention on the sound, on the nuance and character of the confessing voice. As such, these scenes show how the confessing voice can also be distinct from any visual evidence we see on-screen, perhaps more akin to the anonymity of the religious confessional space.

The privatization of the ritual of confession in the Catholic church coincided with a shift in the church's position on the public performance of moral conformity.[4] Sociologists Mike Hepworth and Bryan Turner explain: 'As the church came to emphasize the importance of subjective guilt and the private motivation to contrition, morality was no longer regarded simply as a set of external, objective laws and it was the internal conscience of the private individual which provided the psychological linkage with the requirements of the social order' (1979: 223). This connection between the ritual of private confession, internalized guilt and the clearing of one's own conscience is a key theme in the final episode of *The Jinx*, as well as in the discourse that followed its broadcast. As the documentary comes to a close, Durst breathlessly belches his way through the now-infamous speech, ending with 'What the hell did I do? Killed them all of course'. Critical and popular discourse was quick to label this a confession: 'HBO Documentary "The Jinx" Ends with a Surprising Confession' read the headline from NPR's *All Things Considered* the next morning; entertainment news site *The Wrap* ran with 'Shocking Robert Durst Confession Marks Extraordinary Moment in Documentary History'.[5] Even the film's director interpreted Durst's speech this way, explaining that he believed that Durst had reached out to him in the first place because 'the compulsion to confess is a driver for him. It's a release he was looking for' (O'Connell 2015). But this statement, I would argue, is not as clear-cut as it might seem. In fact, Durst's 'confession' is in many ways a test case for thinking through the significance of the voice for contextualizing the truth value of confessional speech.

The Jinx is true-crime storytelling at its tabloid best. It's lurid, it's suspenseful and it comes with a dramatic narrative conclusion, which was then punctuated in real life by the subject's arrest on the eve of the final episode's broadcast. The film was directed by Andrew Jarecki, whose debut documentary feature *Capturing the Friedmans* (2003) also sensationally explored the vagaries of the criminal justice system by looking back at the convictions of accused child molesters Arnold and Jesse Friedman. Later, Jarecki directed *All Good Things* (2010), a 'thinly veiled dramatization' of Durst's involvement in the disappearance and suspected murder of his first wife, and it was after this film's release that Durst approached the filmmaker about telling his side of the story (Fretts 2015).

The Jinx takes an investigative approach with Durst as its subject. Jarecki draws on interviews, archival footage, police evidence and his own adversarial relationship with his subject to build an argument about Durst's culpability in the death of his wife, as well as a series of subsequent murders. This investigative approach, rhetorical at first, becomes increasingly literal after Jarecki unearths some previously undiscovered material evidence and conceives of a plan to confront Durst with it on camera. At the end of the series' fifth episode, Jarecki finds a handwritten letter that seems to prove incontrovertibly that Durst is guilty of the murder of Susan Berman, a long-time friend who had been killed shortly after becoming a person of interest in his wife's murder investigation. So, in the sixth and final episode of the series, Jarecki confronts Durst with this new evidence, hoping he will implicate himself in the crime.

Despite a few uneasy moments, Durst's reaction is largely anti-climactic after the build-up and planning that had gone into the ambush; Jarecki certainly doesn't get the on-camera confession he was hoping for. Once the interview is over, the crew begins to pack up their equipment and Durst heads to the bathroom, still sporting the lapel mic he wore during the interview. As the final seconds of the documentary wind to a close, we hear Durst talking to himself through this still-hot mic:

> There it is. You're caught. You're right of course. But you can't imagine. Arrest him. I don't know what's in the house. Oh, I want this. What a disaster. [Burp]. He was right. I was wrong. And the burping. [Dry heave]. I'm having difficulty with the question. What the hell did I do? [Burp]. Killed them all of course.

This 30 seconds or so of audio, inadvertently recorded, a seemingly accidental admission of guilt, uttered within a private, off-camera space brought *The Jinx*'s twisted tale of murder, greed and hubris to a shocking, dramatic close. But how are we to characterize this speech? Is it confession? Soliloquy? A cry for help? A fatal slip-up in what had been an almost perfect crime? Recorded off-camera, this disembodied utterance begs the question: Is this evidence, and if so, of what?

I would contend that the reason critics and viewers have been so quick to interpret Durst's speech as confession is rooted in the particular qualities of his speaking voice. The actual words that Durst utters give us an incomplete picture of his culpability, but the breathless quality of his tone and the gaseous eruptions that punctuate his delivery signify something more sinister. Writing about the psychology of confession, Peter Brooks has argued that 'the confessions that seem most authentic to us ... are most often not those freely offered, but those that are in some sense compelled, coerced, wrested from the resistance of the speaker' (Brooks 2005: 74). And it is easy to interpret Durst's voice in this way, as if a valve has been opened and the words that he has long kept hidden are finally being released. His dry heaves, his hiccups, his belching and wheezing are like sulphurous emissions, rising from the belly of a swamp. Moreover, his recognition that the burping is involuntary reinforces his lack of control over the words he is spewing.

Contrast this to one of Richard Kuklinski's many confessions in 'The Iceman Confesses', the second instalment in HBO's three-part documentary series *The Iceman Interviews* (2003). Early in the film, the notorious mafia hit man delivers a clear confession to one of his kills – he not only admits to the murder, but provides relevant details about the crime, along with his own personal feelings about its commission. He reminisces about lighting a man on fire with a smirk on his face and laughter in his voice. Kuklinski's is not a story being wrested from his unconscious, but rather a memory that he looks back on with pride.

The stated motive underlying the production of the *Iceman* documentaries was twofold: filmmakers hoped to uncover details of various unsolved crimes, but they also hoped, in the words of the films' voice-of-God narrator, 'to penetrate the mind of Richard Kuklinski, a mind made for murder'. Confession in this series is pitched as a psychological tool, an act that compels Kuklinski – who has admitted to murdering more than 100 people – to divulge these details as a sort of penance for his actions. But, confession in these films is also a piece of theatre, an act of self-aggrandizement as Kuklinski is given a platform to publicly perform his 'cold-blooded killer' identity. This is in line with Foucault's argument that confession 'is of the order of drama or dramaturgy' (2014: 210). Like David Harris's, Kuklinski's is a voluntary confession, but the boastful quality of his voice situates this as a performance, one that has been exaggerated for impact. As a result, Kuklinski's voice may actually work to subvert the truth value underlying his confessional statements.

So, what if we were to also read Durst's speech in the closing moments of *The Jinx* not as uncontrollable truth-telling but as an act of drama? Some support for this reading comes earlier, in the fourth episode of the series, during a previous interview Jarecki had conducted with Durst. They have been talking about Durst's not-guilty verdict in his trial for the murder of former neighbour, Morris Black. During the break in the formal interview, Durst begins to mumble under his breath, 'I did not knowingly, purposely lie. I did not knowingly, purposely lie. I did not knowingly, purposely, intentionally lie. I did make mistakes'. At this point, his lawyer rushes up to remind him that his mic is on and his voice is still being recorded. Rather than express any disquiet at his words, Durst nonchalantly explains that he'd just been practising how best to explain the situation for the purposes of the film. Looking back on this scene, it is difficult not to read this as evidence to support the intentionality of his words in the later scene. His encounter with the microphone here makes it harder to read his later pronouncement as accident. The gastric tics seem less markers of authenticity than facets of a nervous performance.

Conclusion

Robert Durst was arrested the evening before *The Jinx*'s final episode was broadcast and is (as of July 2018) sitting in a Los Angeles County jail awaiting trial for the murder of Susan Berman. In a series of pretrial motions and statements, he and his lawyers have argued that he was high on methamphetamines during the filming of the documentary (Hamilton and Dolan 2016) and that his history of uttering false or misleading statements should preclude the courts from taking seriously any statement he has made relating to any murders (Bagli 2016). Brendan Dassey's fate is also uncertain at this time. In the weeks following Netflix's release of *Making a Murderer* in late 2015, online petitions calling for Brendan's (and his uncle's) release garnered hundreds of thousands of signatures. In August 2016, a federal judge overturned Brendan's conviction, arguing that the court had 'significant doubts as to the reliability of Dassey's confession'

(*Dassey v. Dittman* 2016: 72). The state of Wisconsin appealed this ruling, but on 22 June 2017, the 7th Circuit of the US Court of Appeals upheld the lower court's decision and ruled by a 2–1 majority that Brendan's confession had indeed been coerced. The Wisconsin Department of Justice has appealed this decision, and Brendan remains in prison. A second season of *Making a Murderer*, as well as a series, *Convicting a Murderer*, that explores the case from the perspective of the law enforcement involved in the case, are both currently in production.

Research has shown that false confessions happen, that the psychological motivations for confession are varied, that police detectives' abilities to distinguish truth from lies are wrong just as often as they are right and that confession is largely overdetermined as compelling evidence in US courts of law (see Kassin 2012). And yet, understanding what these confessions mean, how they could have happened and what their legacy will be seems important to understanding not only what counts as confession, but what confession means right now – in the courts, in documentary and in US culture at large. Both *The Jinx* and *Making a Murderer* captured the cultural zeitgeist upon their release, bringing conversations about true crime and documentary to the forefront of public discourse.[6] Popular press headlines about Robert Durst's confession dominated both arts and crime news circles. Armchair detectives took up Brendan Dassey's cause by looking into the case of Teresa Halbach's murder themselves. When the images and voices of these two individuals were thrust into the public sphere, their words took on a life that exceeded the original context of their utterance, and they were used as fodder for memes, petitions, comedy skits, fundraisers and more.

What do these confessional statements mean for us then, as viewers of these documentaries? Both raise serious ethical questions about the relationship between subject and interrogator – in legal and in documentary circles. Brendan's coercion by detectives and Jarecki's orchestrated ambush of Durst evidence a breakdown in trust and a fundamental lack of faith in the fact-finding process. These were relationships built on both trickery and deceit, where the interrogation process seems to have been designed to elicit words that conformed to a preconceived narrative of the events under question – Brendan as accomplice to murder, Durst as premeditated killer – not a process rooted in evidence gathering and truth-seeking.

What an analysis of voice in documentary brings specifically to the conversation is a mechanism for helping us move past the actual words that were said to focus on how they were said and the context of their utterance.

The voice is a medium, but it is also itself a bearer of meaning. Focusing on voice – in particular, the aural qualities of the confessing voice – can provide insight into how it is that confessional speech becomes evidence which then rises to the level of proof in documentary. Speech, absent the contextualizing evidence of the voice, gives us an incomplete picture of a confession's linguistic meaning. Analysing the relationship between voice and speech allows for a more nuanced understanding of what these confessions mean, which has ramifications for thinking about what exactly it is that these confessions prove. And tracing this connective tissue between voice, speech, evidence and proof is important: for how we come to understand documentary argument, but also for how these documentary arguments operate within the public sphere.

Notes

1 This particular confession was orchestrated by Brendan's attorney, Len Kachinsky, to help secure a plea deal. The film shows Kachinsky and O'Kelly working in concert with state detectives to pressure Brendan into providing corroborating evidence that could be used against his uncle, Steven Avery. While this confessional statement is never used against Brendan in court, it becomes part of the ammunition detectives use to further manipulate his story and elicit subsequent confessions.

2 Some notable examples of these conspicuous public confessions include Bill Clinton's confession to an affair with Monica Lewinsky on national television (1998); Prince Charles's admission of infidelity in a nationally televised documentary (1994); Anthony Weiner's ongoing public admissions to his various sexual foibles (2011–17); Lance Armstrong's confession to steroid abuse in an interview with Oprah Winfrey (2013); and singer Chris Brown's public video apology, uploaded to his own website, acknowledging his assault on a former girlfriend, Rihanna (2009).

3 David Harris was never charged in the death of Officer Robert Wood. However, he was convicted and sentenced to death for an unrelated murder in 1985. Harris died by lethal injection in 2004.

4 Each of the world's major religions has a confessional mechanism, though they 'vary as to how, when, where, and to whom the confessions are given'. The Roman Catholic Church made confession an explicit requirement for all adherents in the thirteenth century (Kassin and Gudjonsson 2004: 35).

5 A number of other publications were also quick to label Durst's speech a confession the day after the final episode's broadcast: 'The Jinx: Robert Durst's Bathroom Confession Is the Creepiest Poem Ever' (*Entertainment Weekly*); 'The Jinx Director

on Murder Confession: “We Just Shook Our Heads”’ (*Rolling Stone*); ‘Is Robert Durst’s “Murder Confession” Admissible in Court?’ (*BBC News*); ‘A Criminal Defense Lawyer Told Us Robert Durst’s “The Jinx” Confession Is “100 Percent Admissible” in Court’ (*Vice*); and ‘Why Robert Durst’s “Killed Them All” Confession Won’t Work in Court’ (*Newsweek*).

6 The first season of *Serial* (2014–), the popular podcast that reopened a decades-old case of potential wrongful conviction to legal and public scrutiny in the fall of 2014, could also be added to this list of true-crime, documentary productions that captured the cultural zeitgeist in the period under discussion. Due in large part to questions the podcast raised, its subject, Adnan Syed, had his conviction vacated and was granted a new trial in June 2017.

References

Bagli, C. (2016), ‘Robert Durst Pleads Not Guilty to 2000 Murder in Los Angeles’, *New York Times*, 7 November. Available online: https://www.nytimes.com/2016/11/08/nyregion/robert-durst-pleads-not-guilty-to2000-murder-in-los-angeles.html (accessed 3 February 2017).

Brooks, P. (2000), *Troubling Confessions: Speaking Guilt in Law and Literature*, Chicago, IL: University of Chicago Press.

Brooks, P. (2005), ‘The Future of Confession’, *Law, Culture and the Humanities*, 1 (1): 53–74.

Chion, M. (1999), *The Voice in Cinema*, trans. C. Gorbman, New York: Columbia University Press.

Dassey v. Dittmann (2016), 14-CV-1310.

Diamond, S., ed. (2010), *Compelling Confessions: The Politics of Personal Disclosure*, Madison and Teaneck, NJ: Fairleigh Dickinson University Press.

Dolar, M. (2006), *The Voice and Nothing More*, Cambridge, MA: The MIT Press.

‘False Confessions or Admissions’, The Innocence Project. Available online: http://www.innocenceproject.org/causes/false-confessions-admissions/ (accessed 3 February 2017).

Foucault, M. (2014), *Wrong-Doing, Truth-Telling: The Function of Avowal in Justice*, trans. S. Sawyer, Chicago, IL: University of Chicago Press.

Fretts, B. (2015), ‘Director of Durst Film Says He Is “Relieved” about Arrest’, *New York Times*, 16 March. Available online: https://artsbeat.blogs.nytimes.com/2015/03/16/the-jinx-ending-robert-durst-andrew-jarecki/ (accessed 3 February 2017).

Hamilton, M. and J. Dolan (2016), ‘Durst Says He Was High in Tapings; Heir Told Los Angeles Prosecutors Last Year He Was on Meth during Interviews for 2015 Miniseries “The Jinx”’, *Los Angeles Times*, 17 December: B1.

Hepworth, M. and B. Turner (1979), 'Confession, Guilt and Responsibility', *British Journal of Law and Society*, 6 (2): 219–234.

The Iceman Interviews (2003), [DVD] Dir. Arthur Ginsberg, USA: HBO.

Kassin, S. (2008), 'Confession Evidence: Commonsense Myths and Misconceptions', *Criminal Justice and Behavior*, 35 (10): 1309–1322.

Kassin, S. (2012), 'Why Confessions Trump Innocence', *American Psychologist*, 67 (6): 431–445.

Kassin, S. and G. Gudjonsson (2004), 'The Psychology of Confessions', *Psychological Science in the Public Interest*, 5(2): 33–67.

Kramer, L. (2014), 'On Voice: An Introduction', in W. Bernhart and L. Kramer (eds), *On Voice*, vii–xv, Amsterdam and New York: Rodopi.

Lassiter, G. D., M. J. Lindberg, S. K. Pinegar, and L. J. Ware (2011), 'Understanding the False-Confession Phenomenon', in S. Diamond (ed.), *Compelling Confessions: The Politics of Personal Disclosure*, 110–129, Madison, NJ: Fairleigh Dickinson University Press.

Making a Murderer (2015), [Film] Dir. Moira Demos and Lucia Ricciardi, USA: Netflix.

McCormick, C. (1972), *Handbook of the Law of Evidence*, 2nd edn, St Paul: West Publishing.

O'Connell, M. (2015), '"The Jinx" Creators Break Silence, Call Robert Durst "Not Just a Random Killer. He's a Strategic Killer"', *The Hollywood Reporter*, 10 June. Available online: http://www.hollywoodreporter.com/news/emmys-jinx-creators-break-silence-801240 (accessed 3 February 2017).

Oregon v. Elstad (1985), 470 U.S. 298.

Payne, L. (2008), *Unsettling Accounts: Neither Truth nor Reconciliation in Confessions of State Violence*, Durham, NC: Duke University Press.

Shipler, D. (2012), *Rights at Risk: The Limits of Liberty in Modern America*, New York: Alfred A. Knopf.

Tell, D. (2012), *Confessional Crises and Cultural Politics in Twentieth-Century America*, University Park, PA: Penn State University Press.

The Thin Blue Line (1988), [Film] Dir. Errol Morris, USA: Miramax.

Wilson v. United States (1986), 162 U.S. 613.

15

Sounds of Disillusionment and Discord: When Pasolini and Sontag Visited Israel

Ohad Landesman

When he was brought to the Jordan River during a trip to Israel in 1970, US Senator Henry Jackson became convinced he was the victim of a hoax. After being assured that it was no joke and that the river was real, Jackson supposedly said that the river's universal distinction, derived directly from its significance to both Judaism and Christianity, 'was an act of public genius'. At another time, with a different distinguished visitor at the same location, Henry Kissinger's immediate impression was that the Jordan River had 'more reputation than water' (Siegel 2015: 222). Whether fictitious or real, such comments made by or attributed to the senator and the secretary of state not only reflect a discrepancy between the imagined shape of Israel's internationally famous river and its disappointingly real proportions, but also attest to how Israel as a place has been perceived and imagined differently by international visitors.

In this chapter I consider two documentaries, made by renowned non-Israeli artists, which chronicle visits to Israel in its early years of existence: Pier Paolo Pasolini's *Scouting for Locations in Palestine* (*Sopralluoghi in Palestina*) (1963) and Susan Sontag's *Promised Lands* (1974). Both of these films, each with a different documentary strategy, offer a retelling of the Zionist dream and represent its unavoidable collapse within a particular historical moment. By exploring the gaps between what Pasolini and Sontag had hoped to see in Israel and the sights and sounds they actually encountered, one can expose not only the tension between expectations and reality, but also the value of an outsider's perspective in revealing the cracks in a monolithic historical discourse. While the films discussed will not be specifically referred to here as 'travelogues' or 'tourist films', as their function is not to simply depict an appealing landscape that entices the audience, it will be shown that they often utilize travelogue elements. They embrace a personal or an essayistic rhetoric that constructs an outsider-looking-in perspective, a position that self-consciously acknowledges the filmmakers' inability to fully and accurately capture the cultures and peoples they are representing. Unlike the tourist film, which often presents a complete alignment between expectations and findings, the films discussed here dramatize the discrepancy between imagining a place and visiting it in reality.

To this end, I will focus in particular on the use of human voice within the sonic landscape of each film. The essential plethora of human voices in both films, I will show, together with the manner in which the voices often ruminate critically on the images they accompany, create and perform a strategic position of outsiderness and provide unique impressions of a place the filmmakers are visiting for the first time. I hope to move beyond Bill Nichols's well-known and important suggestion that the voice in documentary conveys mostly 'a sense of a text's social point of view' (1983: 18) and therefore functions as merely a vocal metaphor to speak about authorship and to reaffirm the filmmaker's point of view. One of the main problems with Nichols's argument, as Trish FitzSimons notes, is that 'it assumes that the voice of the film-maker should *always* be privileged over that of other inputs' (2009: 133, emphasis in original). By showing how the human voice in the films under discussion serves different purposes simultaneously, including an evidentiary function, a practice of the filmmaker's subjectivity and a performance of scepticism, I will side with FitzSimons when she refers to the voice not as a unitary entity but as 'braided', a form of 'stranded singularity' in which 'coming to voice' typically includes the input of many individuals and institutions (FitzSimons 2009: 131). As I will show, Pasolini performs a vocal fluidity, elevating his voice of disillusionment and allowing the sights he encounters in his visit to take precedence over his preliminary fantasies. Sontag, on the other hand, creates a dialectical structure of several voices in order to resolve the problem in documenting a place she simply does not know enough about and uncovers its cultural and historical strata. In *Promised Lands*, the plurality of voices works on several sonic dimensions: opposing two narrators with different points of view, exposing a nationalistic discourse through non-diegetic group singing or directly recording the immediate outbursts of agony from post-traumatic bodies. Sontag also owns up to her intellectual needs by keeping her position removed and non-interventionist, directly observing and listening to a nation in a state of panic and despair. While using FitzSimons's understanding of a 'braided voice' to explain the vocal plurality embedded in the sonic landscape in both films, my discussion will also move beyond the idea of documentary voice as merely a metaphor for a specific point of view. I will simultaneously point to how different material properties in the voices heard (such as changes in tone or volume) move beyond the linguistic meaning of words themselves.

Postcards from the edge: Fantasy and reality in Pasolini's location hunting trip to Israel

In the early summer of 1963, Italian filmmaker Pier Paolo Pasolini embarked on a trip to search for film locations in Israel that would end in utter disappointment. Accompanied by a newsreel photographer and a local Catholic priest named Don Andrea, Pasolini retraced the itinerary of Christ along the border of Israel and Jordan. His travel to a geographic region that he referred to with no specific or relevant political meaning as 'Palestine' was part of a plan to scout locations for *The Gospel According to St. Matthew* (1964), a film Pasolini was hoping to shoot in historical locations from the Bible, such as Nazareth, Jerusalem and Bethlehem. In *Sopralluoghi*, a sketchy, messy and personal unmaking-of documentary that chronicles the director's failed attempts to find appropriate locations in Israel and Jordan, Pasolini reflects on his gradual but growing distress at the unsuitable scenery and the ensuing act of transposition: the impressions he gathered from Palestine would substantiate his later decision to film *The Gospel* in the wild Italian scenery of Calabria instead.

Pasolini's expedition to Israel is characterized by an inevitable intersection between fantasy and reality, romantic expectations and practical findings. Interested in locating archaic forms that remain intact within contemporary surroundings and in identifying grandeur in a 'ruinous landscape of contemporary poverty' (Steimatsky 2003: 240), Pasolini was thwarted. While he was able to locate traces of the poetic and archaic in the fifteen-year-old state of Israel, he also noticed very quickly that filming there would be unsatisfactory due to the modern invasion that had completely altered the biblical settings. Pasolini was searching for leftovers of revolutionary energy from the past, a biblical world that would be preserved and unspoiled, but found out that capitalism had refurnished these locations with new signs and meanings. The modern, industrial scenery he discovered in Israel was, for Pasolini, 'a practical disappointment'. The same place that three years earlier held for Chris Marker a failed promise of social utopia would eventually prevent Pasolini from fulfilling his dream of archaic authenticity.[1]

In *Sopralluoghi* Pasolini and his crew drive across the country to all the major promising locations: Mount Tabor, Christian sites around the Sea of Galilee, a Druze village, an Israeli Kibbutz and finally the Dead Sea, before crossing the border to Jordan. The trip, however, forms for them nothing short of a continuous narrative of disillusionment or, as Ella Shohat describes it, a 'crushing disenchantment with the holy land scenery' (Shohat 2010: 292). Breathtaking images of pre-occupation Israel are juxtaposed with a rapid, semi-improvised commentary that never ceases to directly address the viewer with scepticism. At one point, shortly before Pasolini arrives in Nazareth, he foreshadows what we are about to see next: 'In fact, in a few minutes, as you follow our Fiat's route as it took us towards Nazareth, towards the heart of Galilee, towards Lake Tiberias, you'll see a landscape contaminated by the present.' A few minutes later Pasolini flags the site as a source of dissatisfaction. 'Here it is', he addresses the viewer; 'no further comment is needed to explain why it is totally unusable for our film'. While images of the ancient city do not reveal any visible signs to justify this disenchantment, Pasolini's voice, with its sour tone and downcast quality, shows signs of exhaustion and leads us to the deficiencies of the place by stating that it either has 'too much poverty' or is simply 'excessively modern'. The presence of this disillusioned voice on top of postcard-like images sets up both a 'hierarchy of perception', as Michel Chion refers to it, in which the human voice is accorded privilege over any other sonic elements in the scenery, and a recourse towards an expository mode, in which the meaning inferred by the words spoken overrides any other possible readings that could be suggested by the images (Chion 1999: 6).

The Italian filmmaker, playfully teasing the viewer with his commentary, often talks like a disenchanted tourist, both eager and disappointed to find easy comparisons with sites and sounds from home, between the geography he visits and the more modernized locations in Italy: 'Here we could very well be in Puglia, or in Calabria or Sicily,' he admits. Palestinian towns and villages seem to be wretched, the Jordan River feels like a 'poor, humble, desperate little green river' and the whole territory as imagined by Pasolini is said to be slowly eradicated by sprawling Israeli settlements and industrial plants. On the uniform houses of the Israeli settlers Pasolini utters in dismay: 'You could easily find [them] in the Roman countryside, or in Switzerland.' Such a grim outlook on the sites is embraced right from the beginning of the film, when Pasolini's voice-over addresses the viewer to explain in retrospect why his search in the Middle East ended up being misguided, and why the modern industrial aspect of Israel formed a 'practical disappointment' and a good enough reason to film his Gospel elsewhere. This expedition to the Holy Land is documented and reconstructed like a voyage where the sound humiliates the image, but not in an overtly assertive manner. Pasolini's narration, delivered in a definitive but somehow also embarrassed tone, belittles the impressive settings the images show us so that the visible contradicts the audial, and the real is shamefully obfuscated by the imagined. The land, in his words, forms 'an incredible impression of smallness' and teaches him 'a great lesson in humility'. The Dead Sea is the only location in Israel that makes an impression of grandeur on Pasolini: 'an immense lunar landscape', as he calls it. The rest is simply disappointing in its humbleness and smallness, the unassuming remnants of a once majestic land, so much so that by the end of his quest, Pasolini laments: 'Yes, the biblical world appears, but it resurfaces like wreckage.'

All of Pasolini's encounters with the locals, including his philosophical musings with Don Andrea, are made in the form of cinéma vérité interviews: shot with 16mm equipment, recorded with a Nagra tape and edited mostly in camera. Edgar Morin and Jean Rouch's seminal *Chronique d'un été* (Chronicle of a Summer) (1961), which pioneered this approach, had been released only two years earlier. In the Galilee area Pasolini meets the Druzes, 'Arabs who remained in Israeli territory' and recognizes that they are the only thing in Israel that remained archaic, while the rest of Israeli society has turned quite modern. He hardly speaks to the Druzes, instead layering his contemplative voice-over on top of images of their 'pagan, indifferent, happy and savage' faces. Such presumed authenticity seems far closer, perhaps, to the archaic Gospel characters he is looking for than the new inhabitants of the modern settlements and colonies in Israel. When Pasolini later meets the Bedouin people in the desert their astounding appearance forces him to be explicitly reluctant about using them for his film. 'Their faces are gentle, beautiful, happy', he sadly admits, yet 'a bit sombre'. After all, he reminds the viewer, they are 'pre-Christian'. Admittedly, there is a colonial scent to this location hunting trip, one that Pasolini seems to be very much aware of, but barely able to resist by using his judgemental voice-over to emphasize his authorial control.

The only part in the film where Pasolini speaks extensively and at eye level with local inhabitants occurs when he stops at a kibbutz on the border with Jordan, the same place where 'Christ retreated in meditation'. He interviews an

Israeli family, talking with them about the possibility of fulfilling maternal love in a collectivist structure, and about the unique organization of the kibbutz and its spiritual essence. Interestingly, the conversations are conducted in Italian, not Hebrew, a linguistic strategy that eradicates the local dialect and authenticity of the kibbutz members. It is strikingly also the only part in the film where Pasolini seems at ease and able to create an intimate bond with his interviewees, sitting comfortably on the grass and passing the Nagra microphone between them. The idea of speaking with local inhabitants in his own language about such a nationally specific concept exemplifies Pasolini's position as an outsider, on the one hand, and his wish to avoid the trap of objectifying or misrepresenting his subjects, on the other.

Strangely enough, while this fifty-five-minute film is made in the spirit of cinéma vérité, it can hardly be said to observe and capture life as it is. Pasolini's omnipresent narration is played out serenely and with a tranquillity that the musical score projects as well. Even when Pasolini is directly addressing his subjects on location, his voice is always delivered in a clear and stable intonation that overrides other sonic elements, and few diegetic noises enter the soundtrack. The soothing classical music that accompanies the conclusive tone of narration, arranged almost entirely from Johannes Brahms's compositions, is reminiscent of the dramatic and triumphant score that Luis Buñuel lays over his sarcastic voice-over in *Las Hurdes* (1933). Pasolini, however, is not interested in creating a provocative montage that exploits the tension between a condescending soundtrack and images of misery, like Buñuel is, but, rather, in building a conflict between fantasy and reality, the imagined and the real.

On a practical level, Pasolini becomes utterly disappointed by the end of this trip. 'I found nothing I can use in the film,' he remarks sweepingly, and his ambivalent position towards modernity leads him to ostensibly reject the sites he visits. Don Andrea, on the other hand, understands that the actual and the phantasmatic can also intersect, and therefore do not necessarily need to neutralize each other. Pasolini, so he believes, needs to experience the actual places he visits in order to adapt them elsewhere. The elegant conflict between the two is carefully staged as a discussion between a spiritual mentor (Don Andrea) and a rebellious student (Pasolini). Don Andrea tells him: 'As you walk here you must think, reflect, meditate, to absorb the spirit. Only then can you reinvent it in another place … and then it will become something new.' His reassuring voice often attempts to alleviate Pasolini's overly decisive tone of disappointment, and the pairing creates a tension between an embrace of splendour and the recourse to humility. While Don Andrea's rhetoric is an effort to imagine the decontextualization of a place, Pasolini's is a disillusioned rejection of this romanticism. Pasolini's decision to film *The Gospel* in Italy and not in Israel, a realization that may have occurred long before the end of this trip, is translated in *Sopralluoghi* as not only the culmination of an ongoing vocal struggle between Pasolini and Don Andrea's intellectual positions, but also as the inevitable and practical resolution to a trip that fell short of expectations.

Recording the pain of others: Political dissonance and experimental sound in *Promised Lands*

Promised Lands (1974), Sontag's third film and only documentary, is a small-scale production that took her to Israel during the immediate aftermath of the Yom Kippur War, a decade after Pasolini's visit. Given the filming's proximity to the war and the psychic devastation of its events on the Israeli public, Sontag strolled the streets of Jerusalem and crossed the deserts of Sinai at a challenging time: her visit took place twenty days after fighting began and the conflict was still ongoing. For seven weeks of shooting, with a small crew and marked bravery, Sontag documented a young militarist country confronted by an unexpected crisis and guided by paranoia and fear. She entered the fresh battlefields and drove around the state to closely examine the ways in which Israeli citizens responded to the war. Sontag captured harrowing images of burned tanks and corpses of soldiers surrounded by dried blood and swarming flies in the desert. She also listened carefully to the personal cries of agony and horror, the immediate audial traces of post-trauma.

Sontag uses two different strategies of sonic representations throughout *Promised Lands*: diegetic sound that provides evidence and an essayistic voice-over that opens her film to a rhetoric of questioning and contemplation. She aims at once, as Paul Arthur once phrased it, 'outward to concrete facts and inward to a realm of mercurial reflection', where an 'argument must proceed from one person's set of assumptions, a particular framework of consciousness, rather than from a transparent, collective "We"' (Arthur 2003: 60). Sontag's understanding of film sound as providing indexical evidence correlates with her understanding of photographs as not merely statements about the world but pieces of it, 'miniatures of reality that anyone can make or acquire' (Sontag

1977: 4). Accordingly, Sontag builds a sonic landscape that is composed of unrelated vocal and audial elements recorded on location (prayers, running footsteps, radar beeps, machine-gun fire or radio broadcasts) and edited in rhythmic juxtaposition with abstract images of deserted battlefields, graveyards and supermarkets, open landscapes and clichéd icons of Israeli folklore. The soundtrack is restless and projects anxiety: radio broadcasts are heard on top of interviews, explosions and gunshots intrude upon mourning ceremonies, Arab singing is contrasted with Western pop music, and Muslim and Jewish prayers are heard simultaneously. She records the social fracture of a country confronted by its most dreadful nightmare – forces that work towards its destruction. The sonic topography Sontag captures is composed of a matrix of different categories of voice that not only does not accumulate into one coherent statement or point of view, but also requires careful listening to a diversity of vocal properties whose meaning is derived beyond the words themselves. Sontag forces us to listen to everything that the unsolvable political conflict constitutes in Israel and creates a synthesis that digs further and deeper; she accumulates audial traces, scars of the painful daily reality after the war: the rituals of mourning, the physical pain of wounded soldiers and the mental trauma of Israeli citizens.

At the same time, exploiting the film camera's revelatory powers was not the only function of documentary Sontag had in mind. 'To interpret is to impoverish,' Sontag writes in her 1966 essay 'Against Interpretation', 'to deplete the world – in order to set up a shadow world of meanings. It is to turn *the* world into *this* world. ("This world"! This world, as if there were any other)' (Sontag 1966: 7, emphasis in original). She was reluctant to embrace the conventional form of documentary and this took her to the meditative essay film, where she could engage in open ruminations instead of clear interpretations. Sontag's strategy clearly contrasts with Pasolini's interpretive voice-over that insinuates, in a rather openly declarative personal tone, a certain way to deplete a geography of its archaic values. *Promised Lands* expands on Sontag's reluctance towards didacticism and eschews the use of an expository narration that makes her voice explicit. Since Sontag did not know much about the complex ethnic texture of the young state of Israel and was not familiar with the subtle differences between economic classes, cultural dialects and political groups in the country, the film plays with her inevitable perspective of being an outsider. As an American Jew visiting Israel, Sontag 'comes across not so much as a director, but as a tourist in her own subject' (Hoare 2015), someone who has a vested interest in the place she visits from afar. While she excludes any explicit reference to her own voice in this film, her poetically insinuated critical outlook, expressed through the prism of a visitor, becomes the existing testament by which her carefully reserved view of Israel can be judged. Sontag wishes not to surrender to an Orientalist impulse and simply enforce a foreigner's viewpoint on Israel, but to own up to the fluidity of perspective of a visitor who nevertheless has a vested interest in 'knowing' the place she visits, and consequently has to negotiate the actual and supposed gaps between foreignness and indigeneity.

Accordingly, such personal confusion is performed through the film's main strategy of sound dialectics, where Sontag's point of view is obfuscated but still elegantly coded into the contrast between two distinct male points of view that ruminate thoughtfully from both sides of the political arc: Yuval Ne'eman, an internationally renowned nuclear physicist, speaks about the roots of anti-Semitism and Arab hatred towards the Jewish people, while Yoram Kaniuk, a well-known writer, liberally pontificates about Palestinian rights and the country's dangerous shift from its socialist roots to an American-style commercial culture.[2] The deliberations of these two men run intermittently through the course of the film, both as disembodied voice-overs and on-screen personas, underscoring the 'deepening divisions within Jewish thought over the very question of Palestinian sovereignty'.[3] Sontag, who was influenced by the ideas of Marxist dialectics, surrenders completely to this Hegelian structure of reasoning and produces a rhetorical drama around these two opposing voices, each representing a partial truth. The cinematic drama she constructs aligns with what Bill Nichols identifies as documentaries with 'a more sophisticated grasp of the historical realm', films that 'establish a preferred reading by a textual system that asserts its own voice in contrast to the voices it represents or observes' (Nichols 1983: 648). The recruited voices of Ne'eman and Kaniuk work together with the surrounding diegetic sounds Sontag records and accumulate into a braided and multi-faceted authorial voice that is not singular, but is made of several individuals. In other words, and following Nichols's understanding of voice as a metaphorical sign of authorship, the weight of authority is divided equally between the dual narration and the input of various diegetic voices so that the film 'operates as an autonomous whole, as we do. It is greater than its parts and orchestrates them' (Nichols 1983: 648). The vocal interrelationships in Sontag's work (2003) are a particularly interesting illustration of the 'choric voice', proposed by Trish FitzSimons. Several individual voices provide different inputs and 'aid[s] consideration of documentary voice not as an expression of any single individual, but rather as a collection of braids, albeit often and arguably ideally with a director's perspective forming the main channel at the centre of the braid' (138).

The pairing of two narrating voices exposes the patriarchal underbelly of *Promised Lands*, as its mode of argumentation remains 'trapped' within a Jewish male perspective. Not only are there no interviews with women, no actual presence of the female auteur, and only a few women appear in the film; this elision resonates with Sontag's subsequent claims, made in 2003, that 'war is a man's game' and the 'killing machine has a gender, and it is male' (6). Also missing from Sontag's film are nuanced representations and recordings of Palestinians, rhyming perhaps with Pasolini's decision not to grant Arabs a concrete voice. Excluding one individual shot, where Palestinians are shown crossing the Allenby intersection through Jordan, and not counting Kaniuk's vocal explanation of their suffering, their existence in the film remains 'shadowy and abstract' (Sayre 1974: 44). *Promised Lands* features only Jewish speakers while Arabs remain silent, deprived of a coherent voice and reduced to either exotic scenery (Bedouin herders who become part of a decorative landscape) or nameless bodies torn apart by the atrocities of war.

Where such elements testify to conservativeness, the film's radicalism is felt more strenuously in scenes that deal forthrightly with Zionist ideology. On one occasion, when she visits the wax museum in Tel Aviv, Sontag puts together a montage of bizarre images from the museum that illustrates the Jewish victimhood discourse. On the visual side, she patiently and rhythmically oscillates between various rooms, in which establishing moments in Israeli history are represented in wax – the heroic death of famed soldier Yosef Trumpeldor, the Declaration of Independence, the Eichmann trial or the liberation of the Wailing Wall. As if such pictorial didacticism is not enough, the artificially crafted audio track further points to the specific ideological narrative the museum adheres to. A group singing of 'Shma Israel' (Listen Israel), a prayer that serves as a centrepiece of the morning and evening Jewish prayer services, accompanies the succession of images. It is then followed by another group singing of 'Jerusalem of Gold' (Yerushalayim Shel Zahav), an Israeli song written by Naomi Shemer that is often considered the unofficial anthem of Israel, describing the Jewish people's 2,000-years longing to return to Israel. The synthesis here between image and sound is so blunt and direct that it could only be interpreted ironically. Sontag crafts an implied criticism of the official Zionist discourse in Israel and its limitations. What constitutes collective memory is, for her, an artificially restraining narrative that stipulates dogmatically what is important and what is the ideologically dominant version of Israel's history. Such a critical treatment of national history through group singing is also strikingly different from the individual voices of prayer, mourning and agonizing that Sontag accumulates throughout the rest of the film. The group singing here also functions as a literal demonstration of the choric voice strategy, as described earlier, where a mix of different voices becomes a Greek chorus that comments on reality as one collective voice.

At other times, Sontag carefully expresses her reservations through an edited contrast of individual voices and visual observations. She repeatedly attends funerals and memorial services held only a few days after the war, when the wound is still open and bleeding. The mourning rituals she is filming are forming historical continuity. They are bookended by the opening shots of the film, which show tombstones and funerals of victims from the First World War, and its last sequence, where tanks are making their way to the next conflict. War begets war, Sontag insinuates, and there is no end in sight to this bludgeoned struggle. More specifically, Sontag contrasts in the opening shot a BBC radio broadcast with images from a military funeral of British soldiers. The tombstones, which mostly state 1917 as the year of death, along with the sudden and forceful interruption of the daily broadcast, make an insinuated reference to the Balfour declaration of the same year, a historical mark of colonization and promised British support for a Jewish home in Palestine. A few minutes later, the Syrian minister of Education is heard speaking of the indoctrination of Arab students, reading verses of anti-Semitic hatred from high school books. Such a mediated vocalization of propaganda is accompanied by extremely disturbing images of burned corpses in fresh battlefields. At other moments in the film, such edited vocal associations are only traceable if we listen carefully to the soundtrack; the high-pitched voices of women crying in local funerals are followed by amplified voices of other women praying at the Wailing Wall, thus suggesting an inseparable link between a religious feeling of entitlement and the human price attached to it.

The most memorable, and also most troubling, scene in the film is its penultimate sequence, in which Sontag observes and listens to an experimental treatment for shell-shocked war veterans. In a post-combat rehabilitation clinic, we watch a doctor and a male nurse recreating battle noises of shooting and bombing for a drug-induced patient who seems to be in a state of trance. Banging drawers, slamming beds and shouting orders, they attempt to heal a soldier who suffers from post-traumatic stress disorder (PTSD) using a terrifying treatment of sound re-enactment. Sontag, who patiently observes the situation with penetrating and unflinching direct cinema methods, called the psychiatrist in charge of these therapies (which now seem more torture than therapy) Dr Strangelove. The scene, which certainly belongs to a very specific time period and to treatment principles that may now be totally obsolete, functions as a terrifying cinematic allegory of a haunted society and embodies the feeling of claustrophobia a nation experiences in a tragic moment. As viewers, trapped in Sontag's silent and merciless gaze on this patient twisting in pain and covering his face with a pillow, we are left with nothing else to do but to look at his suffering and listen to both the sounds that generate it and his own indistinct murmuring. The patient, who seems to be in a state of trance from the over-stimulation of sounds, can barely speak and only whimpers occasionally. The artificial sonic landscape created for him, composed of a tape-recorder playing back horrific elements from the battlefield and diegetic sound effects performed by the staff, merges seamlessly with the multi-layered sonic composition Sontag uses throughout the rest of the film. Regarding the pain of others, as Sontag would later title her last published book on war photography, means not only watching it in photographs, but also listening to it on film.

As a frightening testament to masculinity in crisis, the hospital scene was probably also the main reason behind the Israeli government's decision to ban *Promised Lands* in Israel upon its initial release, fearing it would damage the collective national morale.[4] After all, by showing a variety of unresolved complexities, this was a film that was ahead of its time. It neither imitated the nationalistic and heroic cinema made in Israel during the immediate post-Independence era, nor did it align itself with the worldwide wave of support the country was enjoying after the Six-Day War. Her conservative inclinations notwithstanding, what Sontag captured with her film camera – images of a 'beleaguered, paranoid, and terrified nation grappling with the traumas of persecution, war, pain, and death' (Toukan 2012) – was not easy to swallow. Nowadays, the film seems painfully and tragically prescient. It deals with a watershed moment of national rupture in Israeli history, rarely grappled with in either fiction or documentary Israeli cinema (Amos Gitai's *Kippur* (2000) being one striking exception), and provides rare documentation of the outcome of war from the perspective of an outsider. Sontag's film is a pioneering attempt to illustrate the severe moral crisis the country experienced following the war, and the resulting sobering-up from the euphoria of the Six-Day War. As a hybrid of sorts between news reportage and an anthropological essay film, *Promised*

Lands documents the sociopolitical catastrophe in Israel, the inevitable collapse of the Zionist dream as it is eaten away by the reality of a continued Jewish–Arab conflict. Rather than dictate a message, the film asks its audience members to experience, reminding them that 'every image', as Sontag wrote close to the end of her life, is first and foremost 'an invitation to look' (2003: 45). Taking it a formal step further with an ongoing attempt to record the vocalizations that register a traumatic moment, *Promised Lands* is an essayistic struggle that requires not only watching, but also listening to the physical and mental pain emerging from an historical moment of national rupture.

Conclusion

In this chapter I have identified the sonic techniques employed by Pasolini and Sontag in registering, responding and occasionally contributing to the sense of political unrest, scepticism and doubt that characterized their early visits to Israel. I have particularly focused on several permutations of the human voice in achieving this goal, whether as a site for communicating pain and authenticating trauma or a practice of authorial subjectivity that performs a position of foreignness. By embracing different modes of narration and employing international modes of address, especially the essayistic and observational cinema, Pasolini and Sontag offer the distanced gaze of outsiders that is nonetheless suffused with personal stakes in the location they visit. In terms of 'voice', the essayistic provides Sontag an inquisitive mode that is nonetheless marked by her own subjectivity, while the observational allows her to listen carefully to an array of vocalizations and present the viewer with a braided voice that becomes more than the sum of its parts. Pasolini, on the other hand, uses cinéma vérité to assert control with his voice-over and perform a sceptical tone that comments on a geography he had imagined otherwise. At the same time, Pasolini avoids easy didacticism by contrasting his own voice with others, thus formulating self-criticism of his own fantasy of an archaic Israel. Respectively, both Sontag and Pasolini are responding to their foreign perspective by employing different uses of the voice in documentary that testify to such outsiderness.

Furthermore, what makes these films interesting is how they nurture and complicate the gaps created between their filmmakers' initial expectations and what they actually achieved in the final projects. When listening to the vocal and sonic landscape in these two documentary attempts, we notice how it dramatizes the complex and often turbulent relations between envisioning a place from afar and experiencing it in reality. Pasolini and Sontag came to Israel with different aspirations for observation, different legitimizations for their journeys and separate fantasies about the nature of their object of study. While Pasolini's vision of Israel as a reservoir of archaic forms was shattered by the disillusioning traces of capitalism and modernity he came across during his trip, Sontag's curiosity and vested interest in Israel as an American Jew made her practical findings even more distressing than what she had expected. Their fluidity of perspective, channelled in their films through different vocal strategies meant to negotiate their outsider point of view, marked them as foreigners in a place in which they were forced to reassess their fantasies and expectations.

Notes

1 A visit that would be documented in his essay film *Description of a Struggle* (1960). On the importance of Marker's film, along with Sontag's *Promised Lands*, to Israeli film scholarship on the basis of how both films serve as transnational extensions to the cultivation of a national Israeli cinema, see Landesman (forthcoming).

2 Remarks of this kind, it should be noted, were pretty rare at that time in Israel, when the debate about Palestinians' rights was practically non-existent within the public discourse.

3 Taken from the Fandor website's description of the film.

4 The hospital scene resonates in many ways with John Huston's *Let There Be Light* (1946), which provided an unprecedented look into the psychological wounds of the Second World War, specifically PTSD among returning soldiers. While Huston's film was produced by the US army in 1945, it was first publicly screened in December 1980, seven years after Sontag returned from Israel. As a penetrating look at a medical procedure, the hospital scene also echoes Frederick Wiseman's *Titicut Follies* (1967), the first direct and merciless documentation of the casually inhuman hospital treatment of the criminally insane.

References

Arthur, P. (2003), 'Essay Questions: From Alain Resnais to Michael Moore', *Film Comment*, 39 (1): 58–63.

Chion, M. (1999), *The Voice in Cinema*, ed. and. trans. C. Gorbman, New York: Columbia University Press.

Chronicle of a Summer (Chronique d'un été) (1961), [Film] Dir. Edgar Morin and Jean Rouch, France: Argos Films.

Description of a Struggle (Description d'un combat) (1960), [Film] Dir. Chris Marker, Israel/France: Van Leer Productions.

FitzSimons, T. (2009), 'Braided Channels: A Genealogy of the Voice of Documentary', *Studies in Documentary Film*, 3 (2): 131–146.

Hoare, L. (2015), 'Susan Sontag's Panned and Banned Israel Documentary', *Forward*, 8 January 2015. Available online: http://forward.com/schmooze/212245/watch-susan-sontags-panned-and-banned-israel-docu/(accessed 4 November 2017).

Kippur (2000), [Film] Dir. Amos Gitai, Israel/France: Agav Hafakot.

Land without Bread (Las Hurdes) (1933), [Film] Dir. Luis Buñuel, Spain: Ramón Acín.

Landesman, O. (forthcoming), 'Moments of Innocence and Fracture: Fantasy and Reality in Two Documentary Visits to Israel', in R. S. Harris and Dan C. (eds), *Israeli Cinema: Beyond the National*, Minneapolis, MN: Indiana University Press.

Let There Be Light (1946), [Film] Dir. John Huston, USA: U.S. Army Pictorial Services.

Nichols, B. (1983), 'The Voice of Documentary', *Film Quarterly*, 36(3): 17–30.

Promised Lands (1974), [Film] Dir. Susan Sontag, Poland/France.

Sayre, N. (1974), 'Screen: Sontag's "Promised Lands": Treatment of Israel Is at Screening Room', *New York Times*, 12 July: 44.

Scouting for Locations in Palestine (Sopralluoghi in Palestina) (1963), [Film] Dir. Pier Paolo Pasolini, Italy: Arco Film.

Shohat, E. (2010), *Israeli Cinema: East/West and the Politics of Representation*, London: I.B. Tauris.

Siegel, S. (2015), *Let There Be Water: Israel's Solution for a Water-Starved World*, New York: St. Martin's Press.

Sontag, S. (1966), *Against Interpretation and Other Essays*, New York: Penguin.

Sontag, S. (1977), *On Photography*, New York: Straus and Giroux.

Sontag, S. (2003), *Regarding the Pain of Others*, New York: Farrar, Straus and Giroux.

Steimatsky, N. (2003), 'Pasolini on *Terra Sancta*: Towards a Theology of Film', in I. Margulies (ed.), *Rites of Realism: Essays on Corporeal Cinema*, 245–269, Durham, NC: Duke University Press.

The Gospel According to St. Matthew (Il vangelo secondo Matteo) (1964), [Film] Dir. Pier Paolo Pasolini, Italy: Arco Film.

Titicut Follies (1967), [Film] Dir. Frederick Wiseman, USA: Zipporah Films.

Toukan, H. (2012), 'Grappling with Israel: From Sontag to Lacan and the Maoists in between', *Jadaliyya*, 3 September 2012. Available online: http://www.jadaliyya.com/pages/index/7176/grappling-with-israel_from-sontag-to-lacan-and-the (accessed 4 November 2017).

16

Beyond Words: Speech, Ethical Time and the Politics of the Local in Nicolas Philibert's *Retour en Normandie* (2007)

Rhiannon Harries

'At the origin of this film is another', explains the French filmmaker Nicolas Philibert at the outset of his 2007 documentary *Retour en Normandie* (Back to Normandy).[1] That filmic 'other' is René Allio's 1976 historical docudrama, *Moi, Pierre Rivière, ayant égorgé ma mère, mon frére et ma sœur* (I, Pierre Rivière, Having Slaughtered My Mother, My Sister and My Brother), on which Philibert, now one of France's best-known documentarians, began his career as production assistant. The shooting of *Moi, Pierre Rivière* provides the ostensible subject of *Retour en Normandie* – albeit as the first in a back-and-forth relay of historical referents rather than any straightforward 'making of' – as Philibert returns to its Normandy location, re-encountering the rural community from which the earlier film's non-professional cast was drawn and, more obliquely, the historical events they re-enacted. Allio's film was itself based on philosopher Michel Foucault's eponymous dossier, comprising testimony and official legal and medical documents relating to familial murders by the young farmhand Rivière in Normandy in 1835 (Foucault 1973). At the dossier's centre was Rivière's own handwritten testimony, which fascinated Foucault; his lucid textual voice confounded psychiatric diagnoses as well as assumptions correlating his intelligence with his social status and basic formal education.

While Philibert's reference to an 'origin' invokes a sense of time and history through the genealogy of *Retour*, his location of this in 'another' signals the ways in which both will be figured as in excess of his own documentary representation. That this enunciation is made by voice-over, a convention for which the filmmaker has elsewhere expressed his disdain as a remnant of what he calls the 'old, conservative "cinéma du papa"' (Behn 2013: n.p.), hints at an intriguing relationship between temporality, history and this documentary device. Does voice-over here perform 'conservatively', producing impressions of objectivity and certainty, both premised on a linear conception of time, as Bill Nichols suggests in his location of the voice-over commentary at the heart of what he calls the 'expository' mode of non-fiction (1991: 34–38)? Or might the speech of the director work otherwise, interacting with, rather than subsuming, other voices, other moments of speech and non-speech, to create a documentary encounter in which time and history emerge differently? We might expect such questions to find more fertile material in avant-garde documentary contexts, but they perhaps meet with all the more revelatory answers when brought to a filmmaker whose work is often labelled, in slightly dismissive (and notably value-laden) terms, 'old-fashioned' (Bruzzi 2006: 230).

While *Retour* might be where Philibert's work comes closest to the subgenre of 'historical documentary' – understood as pertaining to a more or less distant past relative to the time of a film's inscription – through its engagement with Allio's more overtly historical work of re-enactment of nineteenth-century events, its concern with the contemporary reality of the community in which those episodes took place resists such neat categorization. If, as Nichols claims, documentary realism is itself a 'form of visual historiography' (or, we might emphasize, *audiovisual* historiography) (1991: 177), every documentary is always in some sense a 'historical documentary'. This chapter therefore aims to explore the manner in which voice, speech and language – and, at times, their absences – might participate in the filmic staging of time and history that is always at work beyond that narrow generic label. What part does the voice play in the construction of different kinds of temporality in documentary representations of the historical world? And how might these alternative filmic temporalities modify and complicate the spectator's encounter with that world, with its history and with its other human and non-human participants?

Against the linear

Approaching such questions of time via voice – not least as speech – might seem limited in its possibilities. At various scales, the human voice is associated with linear, successive temporality: the transmission of rational linguistic communication; the child's acquisition of language or what Mladen Dolar sees as the theologically driven teleology of the voice itself ('the voice as the condition of revelation of the Word' (2006: 16)). This linearity is redoubled in the documentary tradition. If the form tends, as Nichols notes, to 'build itself around the spoken word' (1991: 4), voice finds itself predominantly in service to language as a key element in documentary's axial tripartite structure: 'an organising agency that possesses information and knowledge, a text that conveys it, and a subject who will gain it' (1991: 31). Here, the lingering inheritance of European Enlightenment principles leads Nichols to position the form among the 'discourses of sobriety' – systems of knowledge such as science and education – which are driven by rationality, objectivity and, less explicitly, a temporal logic of linear progress. Conventionally premised on the gradual disclosure of knowledge, documentary time recapitulates the Enlightenment notion of history as continual universal progress and perfectibility. Simply put, we watch documentaries with an expectation that in the course of their unfolding we will move towards a position of greater knowledge about their subjects.

Consequently, although the temporal dynamics of voice and speech in Philibert's filmmaking are subtle, the stakes of their resistance to a linear, sequential model of time are high. In displacing the arrow of time and history as progress in *Retour* in order to create a different sense of the relation between past and present moments of the community with which it engages, Philibert cannot avoid undermining the same temporal schema that guarantees the cumulative transmission of knowledge, striking at the epistemological roots of the documentary tradition. Although this chapter limits itself to the study of one filmmaker, it aims to speak more broadly to recent cinematic documentary that fuses conventional observational and expository styles with gently experimental formal choices that avoid, on the one hand, naïve belief in documentary's claim on a prior reality and, on the other, the hermetic post-modern textuality that has lately become an object of critique in view of the so-called post-truth politics.[2] We might place Philibert's work alongside that of documentarians such as Italy's Gianfranco Rosi, France's Raymond Depardon or the United States' Kirsten Johnson, whose practices signal (in ways that include but are not limited to the use of voice and speech) the mediated, partial nature of the knowledge they may offer, but nevertheless maintain an opening onto the world rather than retreating into an airless, hall-of-mirrors reflexivity.

Exploring what we can know about the world and the other beings that inhabit it, and what kinds of relations *not* knowing might enable, opens ethical and political dimensions that this discussion will map with respect to Philibert's work and the place of the voice therein. If we can trace one strand of documentary's genealogy back to Enlightenment thought, we find there a conception of the individual human self structured by knowledge of its environment, a sovereignty which a film such as *Retour* unseats by allowing its filmic subjects to remain beyond the spectator's total comprehension. Illuminating the ethical potential of this spectatorial relation, the readings of *Retour* that follow draw on the thought of French philosopher Emmanuel Levinas, whose interrelated conceptualizations of ethical time and language accommodate possibilities beyond linear temporality and rational signification, providing a suggestive framework within which to consider Philibert's complex mobilization of voices and voicings in *Retour*, in particular the layering of directorial voice-over with bursts of speech and voice from Allio's film and his 2007 interviews with locals, wherein multiple enunciations coexist without smooth chronology or coalescence into a fully thematizable historical narrative. If Levinas's thought helps trace the ethical potential of the dynamic between time and language here, Philibert's documentary practice activates this more concretely through voice, as linguistic communication via speech is invested with the ambiguous temporality of the voice in cinema. Moreover, through its attention to vocalizations and silences that fall at, or beyond, the edge of language, *Retour* implicitly reflects on the fact that only those voices deemed audible as vehicles of rational speech may inscribe themselves in history and the political order. Deploying the French thinker Jacques Rancière's concept of political dissensus as the conflict between who speaks and who does not, this chapter's reading of *Retour* suggests the film develops a mode of audiovisual historiography sensitive to its rural location in which the voices of its inhabitants might become differently audible.

Reclaiming voice-over

Philibert is, in his own words, 'a filmmaker of language' (Philibert n.d.) and there is an almost compulsive return to the exploration of the spoken word – or its absence – across films such as *Le Pays des sourds/In the Land of the Deaf* (1992), in which he installs himself in a class for deaf children, *La Moindre des choses/Every Little Thing* (1997), which centres on the annual performance of a play by residents of a progressive psychiatric institution, and *Nénette* (2010), a portrait of the eponymous orangutan who lives in Paris's Jardin des Plantes zoo. Fascinated by the limits of the human, or rather that which troubles this construct, Philibert's documentaries often seek out subjects – notably non-human animals, but also children and people with disabilities – who are, to differing degrees, socially constructed as other or less than human and politically excluded due to a perceived incapacity for rational speech and, by extension, thought. Clearly, this presents a formal challenge for Philibert, given that documentary's conventional reliance on language reproduces this logocentrism. In response, Philibert makes understated but powerful creative choices that put the spoken word – and especially its operation as documentary voice-over – in question. For example, in *Nénette*, Philibert's camera fixes upon Nénette in her enclosure while we hear the conversations of visitors off screen, inviting the spectator to attend to the gaze and gestures of Nénette and to compare them to the frames of interpretation and explanation that the zoo-goers' commentaries try to impose. Meanwhile, in his most famous film, *Être et avoir/To Be and to Have* (2002), filmed in a one-class rural primary school, the contrast between the child's-eye viewpoint of the camera and the off-screen voice of their teacher as he questions and coaxes the class sets up a variation of the voice-of-God voice-over, recasting it in an interrogative rather than didactic and authoritarian mode. As noted, directorial voice-over is a device that Philibert scrupulously avoids, so its presence in *Retour* is initially puzzling.

As Nichols notes, however, appraising early-twenty-first-century independent documentary, 'it is striking how many recent films rely on the voice of the filmmaker, speaking directly and personally of what he or she has experienced and learned' (2005: 3). He suggests that the spoken voices of (otherwise diverse) filmmakers such as Patricio Guzman and Michael Moore 'maintain their distance from the authoritative tone of corporate media' and that 'their stylistic daring – the urge to stand in intimate relation to a historical moment and to those who populate it – confounds the omniscient commentary of conventional documentary' (2005: 3). Nevertheless, the precise ways in which particular voices and instances of speech in documentary – whether emanating from filmmaker or documentary subjects – operate remain opaque. Elsewhere Nichols elides speech and voice with the sum of formal features of a given film, writing that 'since documentaries are not lectures, questions of speech and voice are not meant entirely literally' (2001: 40) and moves quickly from specific voices towards the concept of the voice *of* documentary, which he holds to be the 'way in which an argument or perspective is expressed […] akin to style […] a distinct form of engagement with the historical world' (2001: 43). This concept of voice enables him to suggest a certain kinship between the work of

Guzman and Moore and Philibert's *Être et avoir*, despite the fact that the latter has no directorial commentary. Although Nichols's piece on directorial voice was written prior to the release of *Retour*, Philibert's inclusion of his own voice in the latter, the signalling of his personal involvement in past and present moments in the history of the community he documents, ought to draw his work further into the ambit of those filmmakers that Nichols sees as using their commentaries in ways that 'speak to power rather than embrace it' (2005: 3). But what exactly separates the speech of these filmmakers from omniscient commentary? How do their voices contest power and avoid partaking of it?

A useful insight arising from Nichols's observations is the implied distinction in the (a)temporal positioning of the voice of conventional documentary's omniscient commentary (the omniscience of which would necessitate a disguising of its temporality) as opposed to the 'intimate relation to a historical moment' that is constructed, at least partly, via the speaking voices of Guzman and others. The ways in which particular voices operate temporally *in* documentary therefore appear key in the creation of the voice *of* a documentary that may either enable or foreclose ethical and political possibilities for the representation of its subjects. Close attention to the temporality of speech and voice in *Retour* may help us to understand how Philibert produces a broader documentary voice that is 'multi-vocal' (O'Donoghue 2011: n.p.) or 'dialogic' (Goldberg 2010: n.p.), doing justice to the ethical time of the Other while also making time for excluded others to be heard.

Enunciating ethical history

If, as Michel Chion writes, in most cinema (perhaps especially documentary) it is the fate of the voice to be forgotten, as the spectator 'retain[s] only the significations it bears' (1999: 1), the first minutes of *Retour* remind us of the materiality if not of the voice as we know it, then at least of vocalization. For four wordless minutes the spectator is plunged into the sensory fullness of a birth, beginning with a close-up of the slippery bodies of newborn piglets writhing on straw beneath the belly of their mother. The low grunts of the sow mingle with the piglets' high-pitched squeaks and as the farmer clips their teeth the piglets emit something close to a human cry of pain, but overall the impression is of a busy soundscape constituted primarily by non-verbal communication. That this scene comes before the pre-credits hints at the ways in which Philibert's film will acknowledge that time and history are never fully contained by any one representation. Its focus on non-human animals might suggest that this is especially the case for those who are assumed to lack not only rational speech but are denied a voice in even the most literal sense (a question to which this chapter returns more fully later in the discussion of non-voice). For the spectator, this is an intriguing opening to Philibert's film.

Levinas's interwoven account of time and language is helpful here for thinking through questions of voice, speech and language whose dynamic in documentary has seemed intractable to precise analysis and with which the opening of *Retour* starts, necessarily obliquely, to engage. Both time and language are refigured in Levinas's proposal of subjectivity as intersubjectivity, in which the Other has priority over the self (the capitalized 'Other' refers to that which is radically different from the self and which therefore cannot become an object of knowledge; this sense of the Other runs through our relations with actual human others). As Sarah Cooper clarifies, 'his is not a developmental theory' but a questioning of subjectivity at its pre-cognitive, primordial foundation (2006: 26). The Levinasian subject is formed in response to the Other, the world is already shared and subjectivity is responsibility to, rather than mastery of, that world and its others. Time and history are likewise shared and relational. Although the lone subject exists in the present and attempts to gather past and future as backward and forward extensions of this present, the past and future in their absolute, ungraspable sense are felt only indirectly through our relationship with the Other. We may consciously live in this 'synchronic' time of the ego, narrativizing time and history into a continuum – an endeavour that the linearity and continuity of conventional documentary time would strengthen – but in our encounters with actual others in the world we find ourselves struck by the interruption of what Levinas calls the 'diachronic', the time of the Other, cutting through this continuity and putting us in contact with a past and future that we cannot know. Dispossessing us of our self-centred temporality, the diachronic expands our ethical relationship to time and history. The past and future are not ours, but we are nevertheless connected to them and made responsible for them through our relations with others. Although indirect, the opening of *Retour* captures something of this dynamic – situated outside the time of the film, the most everyday of farmyard sights is at the same time shot through with the ungraspability of birth, of new lives whose vulnerability (intensified as non-human beings) is palpable.

Following this opening sequence, however, the first sentences of Philibert's commentary return us to the voice-over's familiar domain of linearity and rationality, the synchronic dimension of language as a suppression of otherness that Levinas terms the 'Said'. Over still black-and-white photographs from the shoot of *Moi, Pierre Rivière*, Philibert briefly narrates the genesis of Allio's film via Foucault's dossier and the historical events it documented. His speech serves to soften and mask temporal discontinuities, not only drawing the sequence of still images into the successive flow of narrative but also, as Philibert's account moves from film to dossier to real-life referents, subtly collapsing the distinction between the 1835 episode and its 1976 re-enactment as the indexicality of the photographs of the young man in the role of Pierre (Claude Hébert) is, through language, enlisted in the recounting of the actions of Pierre Rivière himself. Here, in contrast to the sounds of the pigpen, the spectator is unlikely to pay any attention to the voice as medium, leaving it an unproblematized vehicle of language. The voice-over seems engaged in the creation of a sense of time and history as linear and continuous, literalizing Elizabeth Cowie's description of documentary time as a 'present tense – "speaking about the past" – in a "now time"' (2011: 6), mimicking Levinas's description of the ego that attempts to synchronize past and future from its present vantage point. After the pre-credits, Philibert's voice-over continues, now in the manner of first-person account ('I was 24 years old ... Thirty years later, today I have decided to go back to Normandy') installing a strong authorial voice prioritizing selfhood and threatening to make his subjects' experiences mere footnotes in Philibert's personal history, his subjective account belonging no less than the objective, linearity of official history to the Said.

This would seem a disappointingly conventional lapse for Philibert. However, considered via the Levinasian Said, we find within even the most rational linguistic signification an otherness that fissures its unfolding and opens towards diachronic time; a trace of the prelinguistic address of the Other that founds the possibility of communication between the self and that which is beyond it. If, as Michael Renov suggests, Levinas's thought on alterity elucidates documentary's potential access to 'registers of experience not derived from rationalist enquiry' (2004: 149), it is equally significant in pointing to how this experience might take place within and even require the rational domain, given Levinas's claim that the 'betrayal' of the Said is the price for the manifestation the 'Saying' (1981: 7). As Max Goldberg notes, 'unusually for Philibert, *Back to Normandy* is rooted in two levels of direct address: he narrates the film and interviews participants' (2010: n.p.) and it is the interpenetration of these two levels that produces an overarching sense of history as diachronic. Accordingly, if Philibert's voice-over initiates a journey-style narrative that first unfolds along synchronic lines as he undertakes a literal and metaphorical voyage through both space and time to revisit the physical sites of his past, the succession of interviews that follows allows his personal narrative to be displaced by those of others in a diachronic fashion. Sisters Annick and Nicole, for instance, speak emotionally about their involvement in Allio's film as a unique sojourn into a creative world very different from their everyday lives before or since, while elderly couple Annie and Charles implicitly connect the themes of the Rivière affair to the painful account of their daughter's schizophrenia. While some participants speak about specific details from their memories of Allio's shoot, others recount aspects of their life since, such as Nicole, whose description of having to relearn how to speak following serious illness is understandably punctuated by hesitations and filler sounds such as 'euh' ('um') that hover on the edge of communicative speech.

In this way, a multiplicity of partial narratives unfurls; each thread is loosely woven into an expanding diachronic web that proves impossible to map temporally. In the years since the first film, its participants have married; had children and grandchildren; remained single or divorced; suffered grave illness and trauma; built careers; fallen in and out of touch. In their successive recounting, the singularity of these tales of so many lives – and particularly their telling as each speaker approaches their story with different emphases and timescales – makes it increasingly hard to conceive of all these events in the single temporal continuum. Although delivered in verbal language, cinema's audiovisual capturing of these enunciations allows them to be strung together in a way that is not bound to the temporal precision of linguistic grammar; the testimony of the participants necessarily unfolds sequence by sequence, but this does not amount to a successive, synchronic narrative, as each returns to Allio's film as a shared but non-identical moment occasioning their story, suggesting simultaneity and multiplicity instead. The voice-over recounting Philibert's synchronic personal and professional history provides a frame for his subjects' contributions, but is unable to contain them and finds itself continually exceeded by the plurality of history.

At certain moments, the voice-over commentary and its relation to the on- and off-screen presence of Philibert himself reveals that its unsettled temporality is also self-generated, an ambiguty that extends to voice-over in general but which in its more conventional usage is underplayed by filmmakers and unreflected upon by the spectator. Left unconsidered, its effect may be a conjoining of the time of inscription, profilmic and spectatorial time in a synchronizing fashion that initially appears to menace diachronic potential, flattening the difference between the past and present and making history all too graspable. This seems a particular risk at moments where Philibert's voice-over slips into the present tense. 'Today I have decided to go back to Normandy,' over a travelling shot on a French motorway, for instance. 'It's late,' as he studies Allio's production notes beneath a lamp in a dark library. But despite the voice-over's links with certainty and knowledge, it is intrinsically dislocated from the time of the image it accompanies and Philibert also exploits this ambivalence, conditioning our sensitivity to the multiple temporal positions of his voice(s) during the interviews in *Retour* when we hear the filmmaker ask and sometimes answer questions. It would have been easy to edit out his interventions, which depart from his observational style, but the dynamic between his diegetic speech and voice-over – distinguishable by a difference in their auditory qualities that is impossible to articulate in words – begins to unsettle our understanding of, and coincidence with, the filmic present. If we can also hear Philibert talking from within that very present to which the voice-over grammatically attaches itself, when is the 'today' to which the voice-over earlier laid claim? The layering of recordings of Philibert's own voice therefore makes audible a temporal disjunction at the heart of all voice-over. The indexicality of the image and spectatorial assumptions regarding its temporal status as a straightforward matter of pastness relating to a single anterior moment are rendered uncertain by the addition of this voice, which instead aligns itself with the more slippery, shifting meaning of deictic linguistic indices such as 'today' that point to a historical moment but cannot grasp it. In keeping with *Retour*'s broader insistence on the distribution of history through multiple subjects, the synchronization of the voice-over with image and its authority are undermined by Philibert's off-screen voice, which implicates him as no more and no less part of this diachronic structure.

This temporal ambiguity deepens through other voices in *Retour*, not least with regard to archival material from Allio's film, where Philibert further exploits the disjunction between sound and image to disturb the linear relation between past and present that archival imagery often creates. One striking instance occurs at *Retour*'s midpoint, when Philibert films the empty courtroom in which Rivière's trial took place, overlaying his images with snatches of the speech of the actors who played lawyers and doctors in the re-enactment in Allio's film. 'Is Pierre Rivière mad or not?' asks Philibert, posing the question of sanity in the mutually exclusive binary terms of Levinas's Said. Layering the actors' voicing of the contradictory legal and medical assessments, the image of the courtroom is aurally reinscribed with actual historical events and their filmic re-enactment. That these voices are disembodied and multiple renders them strange and spectral, abandoning the realism of the re-enactment in Allio's film to create a temporal indeterminacy. If, as Jaimie Baron argues, the archival in film is most often registered as a 'temporal disparity' by the spectator, 'who perceives a "then" and a "now" generated within a single text' (2012: 106), in this scene the archival voice recordings do not lend themselves to a clearly distinguished model of successive historical moments. Although the spectator learns, via Philibert's voice-over, that the court found Rivière guilty and condemned him to death (this sentence was commuted and Rivière took his own life in prison), the simultaneity of these multiple, conflicting enunciations undermines any clear logical progression to, or grounding of, the judgement. The time of the actual historical events, of Allio's recording of their re-enactment, of Philibert's re-recording of elements of that re-enactment, of the subsequent recording of his voice-over and finally the time of viewing, are impossible to bring into easy chronology in relation to the images, rendering audible the failure of the judgements, of the Said, to achieve closed meaning over time.

While this scene arranges voices with images in ways that create meanings irreducible to linguistic signification, Philibert also enlists voices outside language to intense effect. In a powerful sequence several minutes into the film, having just announced via voice-over his return to Normandy, we see the director's eyes and brow in a tight close-up in his car. Shots of motorway traffic are suddenly intercut with clips of the murder scene from *Moi, Pierre Rivière*, producing a peculiar visual contrast between Allio's nineteenth-century period styling and markers of twenty-first-century modernity. Between images of speeding lorries, we see flashes of the re-enactment of Rivière's gory attack on his mother and siblings with a billhook. Although this might suggest the 'then' and 'now' of Baron's archival effect, sound and above all voice once again render the relation between past and present more complex as we hear the actors' screams. Discussing the function of the human scream in fiction film, Chion emphasizes its narrative placement and describes the 'screaming point' of a film as 'a point of the indeterminate inside the spoken, of unrepresentability inside representation', casting it in temporal terms of suspension, 'a rip in the fabric of time'. (1999: 77) Chion's terms already resonate with the eruption of the Saying within the Said, but Simon Critchley makes the connection to the Levinasian Saying explicit, linking the human scream to the 'affective, pre-linguistic call of the other' (Critchley

1999: 210). The scream as a vocal event thus harbours an affective temporality outside the linear progress of rational language or history, an effect that Philibert amplifies by bridging the scream over the motorway shots and the roar of the traffic over the archival imagery. A more dynamic, unpredictable relation of co-implication between past and present is suggested, in which history would not be a comfortable survey of knowledge, but an exposure to the Other and to actual others, past and future, to whom we are obliged to ethically respond. With regard here to Rivière, whose written testimony all too lucidly described his motives in terms of the poverty of the rural working class – exacerbated by new land-ownership laws – and his mother's financial exploitation of his downtrodden father, this sequence foments reflection on persistent forms of marginalization in non-metropolitan spaces, which this chapter will discuss more fully in the final section.

The dissensual politics of (non-)voice

The screams from Allio's film take part in a chain of interruptive instances of voices beyond the limits of language that includes a number of vocal events featuring non-human animals that work together to gesture towards the fraught politics of voice. The understanding of the scream as an address, without rational signification but replete with the Saying, makes voice a potential nexus at which human and non-human animals might meet and it is notable that Philibert pays particular attention to pigs in the farming environment. As Laura McMahon notes, pigs 'while slaughtered for meat [...] share a striking set of affinities with humans, foregrounded not only by physiological similarities but by levels of intelligence and sociability' (2015: 203). In scenes free of commentary, we hear the first cries of the piglets described earlier, the clamour of a group of pigs awaiting food and, in a difficult scene, the squeal (can we call it a scream?) let out by a pig stunned with a sledgehammer before it is killed. Akira Mizuta Lippit, tracing a line of Western thought that posits animals as incapable of death on the basis of a perceived lack of language, writes that both the animal and cinema, 'haunted by the shadow of death and the absence of language, stake a space beyond the conventions of language and death' (2002: 12). Lippit reads the killing of the giraffe in Chris Marker's *Sans Soleil* (1983) as a trauma that returns insistently and repeatedly from the past, disrupting the temporal logic of the narrative. But Philibert's presentation of the pig's death invites a consideration of voice and language that involves, but is not limited to, trauma and death to the non-human. If Philibert's repeated return to moments where recognizable forms of speech and voice fall away opens diachronic breaks in the synchronic axis of the documentary's unfolding, they also suggest a way in which the interruptive time of the ethical might coincide with the temporality of Rancière's concept of political dissensus, affording us a way of reading the film's construction of time, and the place of the voice therein, as political, despite its distance from overt forms of activist documentary. Rancière claims that the apparently spontaneous constitution of communities as wholes that are divided according to positions and functions is preceded by a general law – the distribution of the sensible – that governs the world of sensory perception, establishing what we can apprehend and parcelling out participation on that basis. Inclusion or exclusion depends on a prior configuration of what can be seen and heard. Exemplary for Rancière is Aristotle's conception of man as a speaking and therefore political animal, alongside the qualification of the slave as understanding but not possessing language, excluding the slave from politics. Dissensus is the essence of the political in its redistribution of the sensible, of what can be seen and heard, when an unrecognized political subject makes an impossible claim on political subjecthood, described as 'the presence of two worlds in one' (Rancière 2001: n.p.). The connection of politics to sensible experience thus allows an understanding of the aesthetic dimensions of documentary – here the organization of time in *Retour* – as politically charged.

While the blow to the pig's head and the cry that it elicits recall the shock of the human screams from *Moi, Pierre Rivière* in the earlier sequence and the eruption of otherness that registers the diachronic time of ethics, the killing of the pig is also woven into a set of moments that attend to a larger problematic of how different enunciations might challenge the Said of political discourse by being rendered sensible beyond language. That the sequence is free of voice-over creates a structural parallel that is reproduced most significantly in a scene in the care home for people with intellectual disabilities where Annick works. That this is a political space is implicitly suggested by the attention the film pays in other scenes to institutional spaces such as the courtroom, the archive where Allio's papers are stored and the prison where Rivière was held. Here, Philibert resists imposing language upon those subjects who have little or none at their disposal, simply filming the residents as they go about their tasks or watch as others do so. The capacity for speech is therefore shown to work beyond human and non-human divides, posing not Lippit's question of how the non-human dies, but how the human or the non-human lives without guarantee of political inclusion. Through deliberate, durational attentiveness to moments of unrecognized voice, or to silence, which the spectator must contemplate without being directed by the rational speech of voice-over, the dissensus that *Retour* performs moves towards a more radical redistribution of the sensible not only by accommodating excluded voices within its audiovisual history, but by simultaneously meditating on the conditions of what is heard as 'voice' at all.

The significance of these instances of unrecognized voices is sharpened in relation to two later sequences which take up broader questions of marginalization and a potential politics of the local that Philibert's project inherits from Allio's insistence on returning history to its proper place. In the first, we see some of the participants from Allio's film organizing and attending a rally against a planned nuclear waste disposal site in the area. In the second, technicians manually cut and join rolls of photographic film at the Éclair laboratory that still holds the print of Allio's film. Philibert notes via voice-over that the founder's great-grandson has just resigned under pressure following a takeover by an outside investment fund and that the 'digital revolution' is now underway. Although oblique, and deliberately without articulation in language, there is an exploration through and across these discontinuous moments of intersecting forms of exclusion that result from the nonlinear temporal logic of capitalism as a process of uneven development, even within a western European country such as France. As scenes of agricultural labour attest, this is not an area in which labour partakes in the digital economy of globalized capitalism, but nor is it a pastoral idyll belonging to the past. Although the screams from Allio's film, the cries of animals and the silences of humans subtly signal the difficulty with which local communities beyond the urban spaces of post-industrial capitalism might contest aspects of globalization, the diachronic temporality that these voices and absences of voice engages also offers a form of resistance to homogeneous global time. If, as Fredric Jameson suggests, late capitalism brings not just the 'end of history' but the end of temporality itself, as a tendency towards a global 'reduction to the present' (2003: 709) of the financial markets, *Retour* insists on the difference of local time and history within the sameness of the global. Constructing an interruptive, layering of times, these voices participate in an unspoken but nonetheless perceptible break with economic and political narratives of progress towards the standardization of time at national and global levels, insisting on the specificity of time and history in local spaces beyond the metropolitan centres of capitalism.

If Nichols is surely correct in his claim that history is what always stands in excess of its documentary representation (1991: 142), it is also the case that documentary in its more mainstream forms works to conceal that margin. Although Philibert's work, as mentioned earlier, does not belong to an obviously experimental tradition of non-fiction, *Retour*'s attention to voices that disturb, rather than guarantee, the linear unfolding of knowledge over time not only makes perceptible the manner in which history exceeds its filmic representation, but also radicalizes it, suggesting that the ethical relation is always an exposure to the otherness of time and history.

While Allio's film sought to recount the Rivière case from within its milieu by casting locals considered by Foucault as 'country people with the same voices, the same accents, the same awkward and unmodulated speech as their counterparts of a century and a half before' (Sheringham 2011: 242), using voice and speech to evince a kind of historical continuity, Philibert's engagement of voice is concerned with how it may put us in contact with others, past and present, while it also effects a break that maintains the otherness of Levinasian diachronic time and the limits of historical knowledge that the latter imposes. In a mode of filmmaking that is intimately connected with information and progressive disclosure over time, often via the linear unfolding of spoken words, Philibert also allows us to reflect on the impasse in which voices, vocalizations and silences to which the film asks us to listen are not heard as such in political discourse, all the while avoiding verbal explanation that would short-circuit this project back to the domain of the synchronic Said. If Nichols endows each documentary with a 'voice' of its own, *Retour* offers compelling possibilities for thinking how the ethical and political potential of that voice depends on its capacity to accommodate the difference of others in and through time.

Notes

1 Translations from the French are my own.
2 On the debate around the complicity between self-reflexivity in post-modern documentary and 'post-truth' or 'post-factual' politics, see, for example, Balsom (2017) and Greene (2016).

References

Balsom, E. (2017), 'The Reality-Based Community', *e-flux* 83, June. Available online: http://www.e-flux.com/journal/83/142332/the-reality-based-community/ (accessed 18 July 2017).

Baron, J. (2012), 'The Archive Effect: Archival Footage as an Experience of Reception', *Projections*, 6(2): 102–120.

Behn, B. (2013), 'Nicolas Philibert's Factual "Fiction"', *Keyframe*, May 31 (accessed 3 September 2015) [No longer available online].

Bruzzi, S. (2006), *New Documentary*, 2nd edn, Abingdon: Routledge.

Chion, M. (1999), *The Voice in Cinema*, trans. C. Gorbman, New York: Columbia University Press.

Cooper, S. (2006), *Selfless Cinema? Ethics and French Documentary*, London: Legenda.

Cowie, E. (2011), *Recording Reality, Desiring the Real*, Minneapolis, MN: University of Minnesota Press.

Critchley, S. (1999), *Ethics – Politics – Subjectivity. Essays on Derrida, Levinas, and Contemporary French Thought*, London: Verso.

Dolar, M. (2006), *A Voice and Nothing More*, Cambridge, MA: MIT Press.

Foucault, M. (1973), *Moi, Pierre Rivière, ayant égorgé ma mère, ma sœur et mon frère -: un cas de parricide au XIXe siècle*, Paris: Gallimard/Julliard.

Goldberg, M. (2010), 'One Village, Two Movies: Nicolas Philibert's Back to Normandy', *Keyframe*, November 17. Available online: https://www.fandor.com/keyframe/transformed-by-the-movies-nicolas-philibert%E2%80%99s-return-to-normandy (accessed 28 December 2016).

Greene, R. (2016), 'The Myth of Authenticity and the Limits of Access: Under the Sun, Cameraperson and the Kiarostami-esque Sublime', *Sight & Sound*, 12 September 2016. Available online: http://www.bfi.org.uk/news-opinion/sight-sound-magazine/comment/unfiction/myth-authenticity-limits-access-under-sun (accessed 28 December 2016).

Jameson, F. (2003 Summer), 'The End of Temporality', *Critical Inquiry*, 29 (4): 695–718.

Levinas, E. (1981), *Otherwise than Being, or, Beyond Essence*, trans. A. Lingis, Pittsburgh, PA: Duquesne University Press.

Lippit, A. M. (2002), 'The Death of an Animal', *Film Quarterly*, 56 (1): 9–22.

McMahon, L. (2015), 'Screening Pigs: Visibility, Materiality and the Production of Species' in M. Lawrence and L. McMahon (eds), *Animal Life and the Moving Image*, London: Palgrave.

Nichols, B. (1991), *Representing Reality: Issues and Concepts in Documentary*, Bloomington, IN: Indiana University Press.

Nichols, B. (2001), *Introduction to Documentary*, Bloomington, IN: Indiana University Press.

Nichols, B. (2005), 'The Time of the Orator', *Film Quarterly*, 59 (2): 3.

O'Donoghue, D. (2011), 'Retour en Normandie', *Senses of Cinema*, October. Available online: http://sensesofcinema.com/2011/cteq/retour-en-normandie/ (accessed 28 August 2015).

Rancière, J. (2001), 'Ten Theses on Politics', trans. D. Panagia and R. Bowlby, *Theory & Event*, 5 (3). Available online: DOI:10.1353/tae.2001.0028 (accessed 20 Jan 2017).

Renov, M. (2004), *The Subject of Documentary*, Minneapolis, MN: University of Minnesota Press.

Sheringham, M. (2011), 'Michel Foucault, Pierre Rivière and the Archival Imaginary', *Comparative Critical Studies*, 8 (2–3): 235–257.

List of Contributors

Chris Cagle is Associate Professor in the Film and Media Arts Department at Temple University. His publications include *Sociology on Film: Postwar Hollywood's Prestige Commodity* (Rutgers University Press), and essays in *Cinema Journal, Screen* and *Quarterly Review of Film and Video*, and a number of edited volumes. His current book project is an examination of an international 'festival film' style in contemporary documentary.

Kristen Fuhs is Assistant Professor of Media Studies in the Department of Communication at Woodbury University. She writes about documentary film, the American criminal justice system and contemporary celebrity, and her work has appeared in journals such as *Cultural Studies, Concentric, The Historical Journal of Film, Radio, and Television*, and the *Journal of Sport & Social Issues*. She is also the co-founder of *Docalogue*, an online space for scholars and filmmakers to engage in conversations about contemporary documentary.

Rhiannon Harries is Research Fellow in Modern Visual Culture at Corpus Christi College, University of Cambridge, where she is also an associate of the Department of French and the Centre for Film and Screen. Her recent PhD thesis explored relations between time, ethics and politics in contemporary European documentary and her current research pursues questions of education and politics in art documentary and contemporary French thought. Her work has appeared in the *New Review of Film and Television Studies* and *The Zoo and Screen Media* (Palgrave 2016).

Brenda Hollweg is Visiting Research Fellow at the School of Fine Art, History of Art and Cultural Studies, University of Leeds. A scholar of Anglo-American literature and culture, she is also a filmmaker. She has published on the cine-essay, particularly on the works of Patricio Guzman, Kathy High, Agnés Varda and Olivier Zuchuat. She is currently co-editing a volume on *World Cinema and the Essay Film* (EUP, with Igor Krstić).

Annabelle Honess Roe is Senior Lecturer in Film Studies at the University of Surrey. She is the author of *Animated Documentary* (Palgrave 2013), which was the recipient of the Society for Animation Studies 2015 McLaren-Lambart award for best book. She is co-editor of *The Animation Studies Handbook* (Bloomsbury 2018) and editor of *Beyond Stop Motion Film: Production, Style and Representation in Aardman Animations.* (I.B. Tauris forthcoming) and has published in journals including the *Journal of British Cinema and Television* and *Animation: An Interdisciplinary Journal.*

Helen Hughes teaches and researches at the University of Surrey. She has published on German cinema, experimental film, GDR and New Austrian documentary. She is the author of *Green Documentary* (Intellect 2014), and co-editor of *Documentary and Disability* (Palgrave Macmillan 2017) and *Deutschland im Spiegel seiner Filme* (CILT 2000).

Sarah Kessler is a media and cultural studies scholar and television critic. She is completing a book on ventriloquism in transatlantic popular culture and writes a regular column on binge viewing for *Public Books*. Her writing has also appeared in the *Brooklyn Rail, Camera Obscura, In These Times, Triple Canopy, Women's Studies Quarterly* and other venues. Kessler received her PhD in comparative literature from the University of California, Irvine in 2016. She currently teaches in the Annenberg School for Communication and Journalism at the University of Southern California.

Ohad Landesman is Teaching Fellow in the Steve Tisch School of Film and Television at Tel Aviv University and a faculty member in the History and Theory Department at Bezalel Academy of Art and Design in Jerusalem. He holds a PhD from the Department of Cinema Studies at New York University. His recent publications appeared in several anthologies and in journals such as *Studies in Documentary Film, Projections: The Journal for Movies and Mind, Animation: An Interdisciplinary Journal* and *Visual Anthropology Review.*

Deborah Matzner is an anthropologist of media and visual culture with geographical expertise in South Asia. She is completing a book based on ethnographic fieldwork with female producers of commercial Hindi-language television and activist documentary film in Mumbai. A member of the board of the Society for Visual Anthropology, Matzner has a background in the making and analysis of ethnographic film. Her latest work seeks to integrate anthropological approaches to media and the senses to attend to the sonic aspects of audiovisual media in India and beyond.

James Nicholson began working in the New Zealand film industry during the 1980s. An experienced director, cinematographer and editor of fiction and non-fiction productions, he completed his MA at Auckland University in 2014, focusing on sound in documentary. He is Curriculum Leader of Television and Screen Production at Auckland University of Technology's School of Communication Studies.

Maria Pramaggiore is Professor of Media Studies and Dean of Graduate Studies at Maynooth University in Co Kildare, Ireland. In addition to publishing three monographs and more than thirty articles and book chapters, she co-authored *Film: A Critical Introduction* (4th edition 2019) with Tom Wallis and co-edited *RePresenting Bisexualities* (1996) with Donald E. Hall. Her work ranges in subject matter from Irish cinema to feminist experimental cinema to the Brony fandom. She is currently co-editing a special issue of *Alphaville: A Journal of Film and Screen Media* on Queer Media Temporalities and a book on AIDS in Irish Media and Culture with Páraic Kerrigan.

Pooja Rangan is Assistant Professor of English in Film and Media Studies at Amherst College. Her book *Immediations: The Humanitarian Impulse in Documentary* (Duke UP 2017) examines the humanitarian ethic of giving voice to the voiceless in contemporary participatory media interventions. Rangan's work has been published in *Feminist Media Histories, Film Quarterly, Camera Obscura, World Picture* and *differences*, among other venues. Rangan also serves on the board of the Flaherty Film Seminar, and is co-editor, with Genevieve Yue, of a special issue of the journal *Discourse* on 'Documentary Audibilities'.

Luke Robinson is Lecturer in Film Studies, Department of Media and Film, University of Sussex, UK. He is the author of *Independent Chinese Documentary: From the Studio to the Street* (Palgrave Macmillan 2013), and the editor, with Chris Berry, of *Chinese Film Festivals: Sites of Translation* (Palgrave Macmillan 2017). His writing has appeared in books and journals including *DV-Made China: Digital Subjects and Social Transformations after Independent Film, The New Chinese Documentary Movement: For the Public Record, The Routledge Companion to Cinema and Politics* and *Journal of Chinese Cinemas*.

Patrik Sjöberg is Assistant Professor in Cinema Studies at Karlstad University in Sweden. He is the author of *The World in Pieces – A Study on Compilation Film*, (Aura Förlag 2001) and a contributor to several anthologies on documentary culture as well as articles and essays in Swedish and English. His research seeks out the shared space between documentary film and other related fields, in particular the avant-garde and experimental film, sound art and media culture. He is currently working on a book on speaking subjects in documentary media and sound art.

Jean Walton teaches English and Film/Media at the University of Rhode Island and is the author of *Fair Sex, Savage Dreams: Race, Psychoanalysis, Sexual Difference* (Duke 2001). Her book *Mudflat Dreaming: Waterfront Battles and the Squatters Who Fought Them in 1970s Vancouver* is forthcoming from New Star Books (2018) and her book *Buffalo Trace: A Threefold Vibration* (co-written with Mary Cappello and James Morrison) is forthcoming from Spuyten Duyvil Press (2018).

Shilyh Warren is Associate Professor of Film and Aesthetic Studies at the University of Texas at Dallas. Her book project reconsiders women's documentary filmmaking of the 1970s. Her essays on documentary and feminist filmmaking have appeared in *SAQ*, *Camera Obscura*, *Signs*, *Jump Cut*, *Mediascape* and *Films for the Feminist Classroom*.

Index

Note: Locators with letter 'n' refer to notes.

CPSIA information can be obtained
at www.ICGtesting.com
Printed in the USA
LVHW091526040619
619991LV00008BA/161/P

9 781501 331251